Order this book online at www.trafford.com
or email orders@trafford.com

Most Trafford titles are also available at major online book retailers.

Print information available on the last page.

ISBN: 978-1-4907-6506-8 (sc)
ISBN: 978-1-4907-6507-5 (hc)
ISBN: 978-1-4907-6508-2 (e)

Library of Congress Control Number: 2015914567

Trafford rev. 09/04/2015

 www.trafford.com

North America & international
toll-free: 1 888 232 4444 (USA & Canada)
fax: 812 355 4082

Nature's twelve

Corners of the Earth

Also by Louis Komzsik

Wheels in the sky, keep on turning

Cycles of time, from infinity to eternity

Gravity's mysteries, from ether to dark matter

Three of life, the perfect number

World of five, the universal number

Magnificent seven, the happy number

To those who see them

Contents

Acknowledgments

I am very grateful to Thomas Flöck for his participation in research and many valuable comments, to Paul Sicking whose recommendations significantly enhanced the style of the presentation, and to Olivier Schreiber for his eagle-eyed corrections greatly improving the quality of the book.

The image of the globe on the cover is by courtesy and permission of its author, Gene Keyes.

I also appreciate the contributions of the staff at Trafford Publishing: Reggie Adams coordinator, and the designer team.

August 29, 2015

Louis Komzsik

1

Prologue

Many humans are fascinated by numbers! It is unknown whether you dear reader also are, but the fact that you opened this book is already promising. This fascination may be rooted in childhood when we first learned to count to three, five, seven and even twelve. The last was a spectacular feat as the fingers had to be abandoned to achieve it.

The title states something that everybody knows: twelve is pervasive in nature. Then the subtitle indicates something that does not exist in reality as we all know. Earth is round, well, spherical. Clearly some kind of an analogy is at work here. We will try to connect these two statements in this book.

The first chapters explain the origins of our penchant for the number twelve and its copious occurrences in the everyday life. This is followed by several manifestations of the number in geometry. Algebraic peculiarities and the confounding symmetry of twelve will be described. Finally, nature's plentiful uses of twelve and the corners of the Earth will be revealed.

This book is an inquiry into finding out whether natural phenomena follow some numbers intrinsically or the intriguing natural occurrences of special numbers are purely coincidental.

4

1

Twelves of time

Human preference for and preoccupation with the number twelve dates back millennia. While there are many competing hypotheses about the origin of this, the most plausible of them is clearly in astronomy. Ancient human societies observed the cycles of the Moon that were very visible on the short term, and also the cycles of the Sun on the longer term. They soon arrived at the conclusion that roughly twelve of the short term cycles constitute one of the longer term. Since those cycles were very important in the lives of humans, the particular number associated with them also became very important.

Many millennia ago Chinese astronomers (well, people looking at the skies whatever their official name was at that time) already created a yearly calendar in twelve parts. They also realized that it did not exactly match the yearly cycle and occasionally an additional short period (let's use our modern term, month) was added to catch up with the annual cycle. They were also aware of an even bigger cycle related to the precession that led them to another use of twelve as we will see shortly.

About the same time, the ancient Babylonians also recognized that the year consists of twelve lunar months.

Their long term observation led to the recognition of a more accurate duration of the lunar month being about 29.5 days. They noticed that 19 solar years is exactly 235 lunar months, that is 19 lunar years plus 7 lunar months ($19 \cdot 12 + 7 = 235$). Hence they could live with a lunar calendar with twelve months for 19 years and then add 7 lunar months to get back in synchrony with the Sun.

That would have made for a prolonged discrepancy with the solar activities. Therefore, in a solution similar to our current leap years, they added an extra month at the end of the 3rd, 6th, 8th, 11th, 14th, 16th and 19th year of the 19 year cycle. This smart solution resulted in a calendar that was never more than 20 days out of synchrony with the solar cycle. In any case, the fact that their fundamental number was also twelve is relevant to our story.

Some other cultures, notably the Romans, were not as quick to establish their twelve-based calendar. The ancient Roman calendar, per anecdotal evidence established by Romulus, the founder of Rome, had only ten months originally. The names of the ten months were: Martius, Aprilis, Maius, Iunius, Quintilis, Sextilis, September, October, November and December.

It is noticeable that some months follow the Latin name of their numeral order, while some do not. Quintilis, Sextilis, September, October, November and December verbatim mean the 5th, 6th, 7th, 8th, 9th and the 10th months. The other months were named after gods and goddesses. Martius was named for the god of war, Mars, Maius and Iunius were for Roman goddesses. Aprilis was designated to honor Aphrodite, the Greek version of Venus, the goddess of beauty.

The ten month long calendar with 29 or 30 day long months was extremely short, less than 300 days. One of the Roman rulers in the 7th or 8th century BC attempted to adhere to the lunar cycles by adding two additional months, Ianuarius and Februarius, to the end of the year and the calendar became 355 days long. It was still short, hence Julius Caesar in 46 BC added several days, as well as moved Ianuarius and Februarius to the beginning of the year, where we have them now. Ultimately in all classic cultures, twelve became the basis of human timekeeping.

There was, however, another celestial occurrence of twelve, related to the aforementioned precession. The precession phenomenon is due to the fact that the axis of Earth is not perpendicular to the plane in which Earth is rotating around the Sun. This misalignment of the axis means that the Earth in its yearly cycle does not return to the same position with respect to the sky. In fact it takes twelve times 2,160, or 25,920 years to complete this humongous cycle, also known as celestial great year. We are part of an even larger cycle, the 230 million year cycle of our solar system around the whole Milky Way which we will omit here.

Looking into the Milky Way ancient human observers recognized that the Sun is traveling around the distant objects in the sky in a repeatable manner. Hence they designated twelve equal sections of the sky with a notable star or constellation in each sector, the famed Zodiac. Those stars were so far that their movement

was barely detectable, hence were considered fixed.
The name's origin is based on the Greek words for
animal circle because many of the dominant constel-
lations in each sector were associated with an animal.
The twelve names were the Ram, Bull, Twins, Crab,
Lion, Maiden, Scales, Scorpion, Archer, Goat, Water-
bearer and the Fish. There are depictions of them by
the Egyptians as early as 3000 BC. Due to the precession, the signs of Zodiac have been shifted since that
time and people born in the Aries sign are actually
born in the Taurus section.

Ancient Chinese, Babylonians, Egyptians, Greeks
and Romans all had their own names and philosophies
related to the Zodiac. Incidentally, most of them associated the same animal with the same constellation.
For example, the Bull (constellation Taurus) appears
in the ancient mythologies ranging from India though
Babylonia and even Europe. The philosophical underpinning everywhere was the Bull being the symbol of
strength. We will not explore further the tremendous
amount of beliefs, influence and conceptual interpretation of the Zodiac. Suffice to state that twelve is its
fundamental number.

After the universal twelve of the Zodiac and the
global twelve of the year, it is reasonable to seek it on
a smaller scale. That is, of course, the day. The Babylonians may have developed their twelve hour days because they had a base 60 system of arithmetic and that
is well divisible by 12. They certainly recognized the
fact that the length of the day varied in the big yearly
cycle, so their measurements may have been rather
loosely interpreted.

The Egyptians also used a twelve hour day according to archaeological records and the prevailing theory
about their arrival at it is based on finger counting.
Since each finger (not counting the thumb) has 3 pha-
langes, it was natural to count the hours by those.
So goes the theory. However, they were also acute
observers of the night sky and another historical ex-
planation is that Egyptians first divided the night into
twelve periods. That was based on the observation
that during the summer nights, between sunset and
sunrise, ten so-called decan stars rose in approximately
hourly intervals over Egypt. This, with one extra hour
for the two twilight periods resulted in twelve. There
are also archaeological records of sundials (for day-
time) and water clocks (for night time) with twelve
notches.

Then there is the story about a pharaoic advisor
named Thedontus who supposedly had a birth defect
of having extra fingers on his hands, called polydactyly.
He proposed to divide the day and night accordingly
into twelve segments. Whether this has any historical
foundation is unclear, but in any case, from the pha-
langes of a finger to extra fingers, we now have several
possibilities to explain the smallest twelve based scale
of humankind: hours.

The Romans also divided the day into 12 hours although their counting was sort of backwards. They
counted the morning hours with respect to the high
noon (meridian). For them 5 ante meridiem meant 7
in the morning in our current usage. We did inherit,
however, their reference notation with AM (ante meri-

diem) and PM (post meridiem), albeit 5 AM is now
two hours earlier than in Roman times.

In any case, most clocks nowadays have an ana-
log face divided into 12 hours with one arm rotating
around twice a day, another arm two time twelve times
a day and a third arm two time twelve time 60 times
a day. This brings the related question: why did we
not end up with subdividing a circle into 12 segments
in the geometrical sense?

This is likely due to the Babylonians' sexasegimal
(base 60) system. While the Zodiac was adequately
subdivided into 12 segments, for precise geometric activity in the circle, a famous pastime of Greek geometers, one needed a more detailed discretization. Eratosthenes, the Greek astronomer living in Alexandria,
a Greek territory at the time, in the 3rd century BC,
divided the circle into 60 partitions. He did so in order
to aid his calculations to determine the radius of Earth,
and he created the first systematic method of presenting the Earth by geometrical means: he invented the
latitude and longitudinal lines of the Earth.

Eratosthenes' latitude lines were not equally spaced
though. They ran through some known places at the
time, such as Alexandria, the place that helped him
make an astonishingly accurate measurement of the
radius of the Earth, but that is another story. His
center of the Earth fittingly was in Alexandria and
he placed both a longitude (meridian) and a latitude
(parallel) line through it. The next meridians to the
west were one at Carthage and another one at the Pillars of Hercules, a hero in our next chapter, which was

the ancient name of the Straits of Gibraltar. The second and third meridians to the east of Alexandria were placed at the Indus and the Ganges, whose presence he was well aware of.

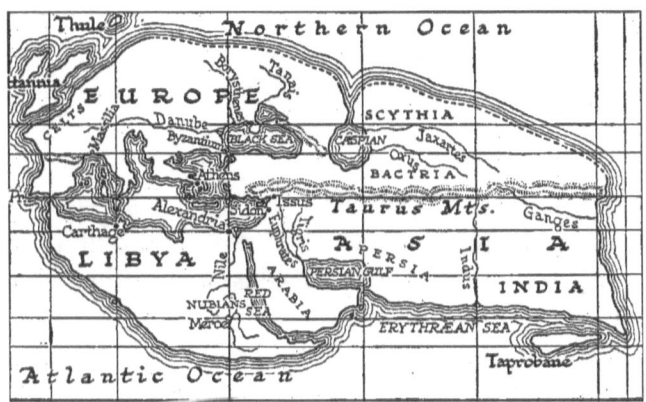

Eratosthenes' map, shown above, is worthy of deeper investigation as it reflects the known Earth around 200 BC. Notice his knowledge about the main waterways, Nile, Danube, Tigris, Euphrates, Indus and Ganges, that are all shown with their modern names recognizable. The Mediterranean coast was rather accurate and the Black Sea, the Caspian Sea, Red Sea and Persian Gulf were also correctly placed. The British Isles were also well defined, but otherwise the continental boundaries were rather approximate.

A half a century or so later, his countryman, Hipparchus, upped him by subdividing each segment of Eratosthenes into six sub segments and generalized the longitude lines to be equidistant rather than connected with specifically chosen and known places. Our most commonly recognized image of the net over the Earth had been finalized more than two millennia ago. In the last chapter we will revisit this topic by applying some twist to it and contorting our Earth into the shape of having corners.

The location of the prime meridian has a history of its own. Ptolemy also produced a world map in his 1st century AD work where he assigned the prime meridian to the area of today's Canary Islands. While he used 180 degrees in both directions in longitude as we do it today, he measured the latitudes in hours upwards from the Equator.

Even Mercator, the 16th century Flemish cartographer, whose projection is still used in today's maps of the globe, used a different prime meridian through the Cape Verde Islands also off the African coast. Ultimately, the zero meridian became the meridian going through Greenwich in the United Kingdom and the location of zero Universal Time.

The different hour long time zones around the world are designated in reference to this as either negative offsets, meaning behind the universal time, or positive offsets for time zones ahead of it. There are some minor deviations from this system to accommodate political entities wider than a zone by locally modifying the straight zone boundaries, but by and large our time measurement around the globe is based on twelve, justifying the title of this chapter.

Before we conclude this chapter we also have to acknowledge that twelve's astronomical emergence and consequent use in our time measurements is somewhat serendipitous. The Moon becomes more distant from Earth by about 4 centimeters, or about an inch and a half, every year. That is only about 8 ten thousandth of a billionth of the considerable 240,000 miles between us, but precipitates a dire future of losing our Moon in a few billion years.

At the same time the rotation of Earth slows by one and a half millionth of a second each year, even today. Had we started to do such observations 600 million years earlier we would have found that Earth's day was about 21 hours. The perilous rotational dance between our Moon and Earth was different and the number of times the lunar cycle repeated itself in the solar cycle would have been closer to 13.

Hence it appears that twelve might be an accidental number, but even if that is so, it does not eliminate the fact that it is now a part of life, ours and the gods as well.

2

Gods of twelve

Having established the likely reasons for distinguishing twelve, let us see how we folded it into our belief system. We must start with the most well known, the ancient Greek mythology. The story starts with the Titans, who were notably twelve. The Titans were the children of the Earth and the Sky, named Gaia (the female) and Uranus (the male).

They were six males, Kronos, Krios, Koios, Hyperion, Oceanus and Iapetos. They were the fathers of the time, the heavens, the intelligence, the light (of the Sun and Moon), the water (of the oceans), respectively. Iapetos was, simply but most importantly, the father of Atlas, who held Earth on his shoulders, and Prometheus, who brought fire to humans and was punished by a vulture gnawing at his liver for eternity.

The six female Titans where Rhea (wife of Kronos), Thea (the wife of Hyperion and hence the mother of the Sun and Moon), Thetys (wife of Oceanus and the mother of the oceans), Phoebe the Titan of the Moon, Mnemosyne, the Titan of memory, and Themys, the Titan of justice.

These twelve have a certain extraterrestrial connection, being the fathers and mothers of components of our solar system. There is also a strong philosophical underpinning of human evolution in heralding intelligence, memory and justice.

But their story turned sour soon. The oldest of them was Kronos who ultimately took power by killing his father, Uranus, a scenario often repeated in mythology. He fathered the gods of Olympus with Rhea. However, he proceeded to eat them after they were born because he was afraid of history repeating itself and his children overtaking him. After Zeus was born, Rhea tricked Kronos into swallowing a rock wrapped in baby blankets instead of Zeus.

When Zeus grew up he did turn against his father. Using a magic potion created by his grandmother Gaia, he forced Kronos to vomit up his brothers and sisters. After that he started the War of the Titans, and gaining victory, he banished the Titans to the underworld. Since Atlas was fighting alongside his father, the Titan Iapetos, he was punished by Zeus to hold Earth on his shoulder, for eternity.

Zeus married his sister Hera and they sit atop Mount Olympus along with the other ten of the twelve gods: Poseidon, Athena, Apollo, Ares, Artemis, Aphrodite, Demeter, Hermes, Hephaestus and Dionysus. Poseidon ruled the seas, Athena was the goddess of wisdom and Apollo the god of arts. Ares was the god of war and Artemis the goddess of hunting. The goddess of beauty was Aphrodite, Demeter the goddess of fertility, Hermes was the messenger and Hephaestus the blacksmith of the gods. Finally, Dionysus was the god of wine and good times.

There were some additional deities, called demigods, such as Heracles and Hades, but not included in the twelve. These two share a notable twelve related relationship worthy of mention: Heracles' so-called twelve labours. According to the myth, Heracles in his madness killed his wife and his six sons (another common occurrence in Greek mythology). His penance for this, ordered by the Oracle of Delphi, was to serve King Eurystheus for twelve years who assigned twelve difficult tasks to him. These are known as Hercules' twelve labors after the Roman name of Heracles.

His tasks were related to a collection of mythical beasts or valuable rare animals, either to kill them or to capture them and bring them back to King Eurystheus. Heracles had to kill the man-eating Nemean Lion, the Lernean Hydra, a water serpent with nine heads and the Erymanthean Boar, an aggressive wild pig. He also had to capture the Hind of Ceryneia, a deer with golden horns and bronze hoofs, the Cretan Bull, given by Poseidon to King Minos to sacrifice but who let it free, and capture the man-eating Horses of the King Diomedes.

Heracles also had some mundane tasks, such as of cleaning up King Augeas' Stables occupied by thousands of cows, goats and sheep, and driving away the Stymphalian Birds, mythical species that supposedly ate humans. Then he had to bring back the Belt of Hippolyte, the Queen of the Amazon women, who got it from Ares, the god of war, the red colored Cattle of the monster Geryon with three heads and pairs of legs, and the golden Apples of the nymphs, named

Hesperides. These were the daughters of Atlas who apparently still had time to sire them while holding Earth on his shoulders.

The name Pillars of Hercules for the Straits of Gibraltar was mentioned in the last chapter as meriting a meridian on Eratosthenes' map. The name's origin is in the story of the monster Geryon who lived on the island of Erythia beyond the boundary of Europe and Africa. According to the legend, Hercules had to split the mountain into two to get to the island, hence creating the Straits of Gibraltar. Geryon's herd of red cattle was guarded by Orthus, a vicious hound with two heads, that Heracles had to defeat to be able to get to the cattle.

Finally, he had to kill Cerberus, the mythical three headed hound guarding the underworld of Hades. Incidentally, Cerberus was the brother of Orthus of Erythia, likely the older as having more heads. Heracles went there and Hades, the ruler of the underworld, agreed to let him fight the creature, but only barehanded without weapons. Heracles subdued the hound with his bare hands and carried it back on his shoulders to King Eurystheus. Ultimately Cerberus was freed and returned to guard the underworld for eternity, one of the few lucky creatures to survive encountering Heracles.

These stories still fascinate humans so much so that a recent two pound coin of Gibraltar even commemorates the labors. The coin has a two headed hound on its reverse with the name of Hercules that is the Roman version of Heracles. On the other hand, the dog's name written on the coin is Cerberus, but Cerberus had three heads. It appears there is some confusion about who had how many heads. But the coin is very pretty and is in use as shown below.

The Romans had the same twelve main gods with different names but identical characteristics and roles. They were in the same order Jupiter, Juno, Neptune, Minerva, Apollo, Mars, Diana, Venus, Ceres, Mercury, Vulcan and Bacchus. Their extra underworld god of Hades was named Pluto. The Roman versions might be better known nowadays, but the original credit goes to the Greeks whose archaeological remnants depicting their gods date back to millennia.

The Romans also carried the number twelve into their first system of law established some time in the 5th century BC. The Twelve Tables, or Lex Duodecim Tabularum, described the first structured set of laws in Rome. The main reason for declaring them was to ensure public awareness, to use a contemporary word, to make citizens aware of the law before they violated it, a fair concept. The code was compiled by a ten-member group called a Decemvirate and originally contained only ten laws but later augmented with two specific supplements.

The Twelve Tables laws were finished around 450 BC, so two and a half millennia ago. The subjects of the tablets were: courts, trials, debt, rights of fathers, guardianships and inheritance, possessions, land ownership, injury laws, public law, sacred law, and the two supplements: marriage law and election law. Whether the augmentation to twelve was to really add missed sections or to make it a round twelve is not known.

The Twelve Tables were instrumental in Romans building a successful society. The average Roman citizen's rights were well defined and social conflicts minimized. They produced the first set of legal guidelines that were soon followed by other ancient societies. Furthermore, their impact to humanity in general is also undeniable as some remnants of the Twelve Table laws still exist in modern civil laws today.

Nordic mythology also contains twelve gods. They

were Odin, Thor, Tyr, Ull, Balder, Bragi, Forseti, Frey, Heindall, Iord, Vali and Vidar. Some of them were the sons of Odin, the head god and Frigga, his wife. The better known are Thor, the god of thunder, Iord, the god of seafarers, and Vidar who is known for avenging his father Odin. Odin and the Nordic gods also had some animal companions, numbered twelve by some accounts: wolves, ravens, and even goats that pulled Thor's chariot.

About the tenth century BC came the first texts of the Hebrew Bible. While it was a work in progress over the next millennium, very early in its development the twelve tribes of Israel appeared. According to the ancient writings, Jacob had twelve sons and a daughter. Jacob's favorite son was Joseph and the other sons were so jealous of him that they sold him into slavery. This ultimately led to the move of the Hebrews to Egypt as written in the Old Testament.

According to biblical records there were twelve tribes or regions of Israel, and the tribes were named after the sons of Jacob, except for Levi who was not designated as a tribe leader. On the other hand, Joseph's two sons both became heads of tribes. The names of the original tribal leaders were: Ephraim and Manasseh (Joseph's sons), Benjamin, Judah, Issachar, Zebulon, Reuben, Simeon, Gad, Dan, Asher and Naphtali.

The next step in chronology is Christianity with its bewildering number of occurrences of twelve. The most well known is, of course, the twelve disciples of Jesus. Their names were Peter, Andrew, James, John, Philip, Nathanael, Matthew, Thomas, James (the Lesser), Simon, Thaddeus and Judas. Their family relationships are well documented: Peter and Andrew were brothers, so were James and John. Their nick-names also entered everyday usage, like doubting Thomas or Judas the traitor, although the latter is being challenged by modern Bible researchers as a misinterpretation of his role on that faithful day described in the Bible.

The New Testament contains numerous references to the number twelve. There is the story of a woman cured by Jesus, after suffering from a hemorrhaging sickness for twelve years. It is also written that the kingdom of God had twelve gates that were guarded by twelve angels. Jesus was twelve years old when, separated from his parents, he ended up speaking to the scholars in the temple. He also resurrected the daughter of a believer who was twelve years old. Finally, Jesus fed thousands of men, women and children with loaves of bread and fish, after which twelve baskets of food still remained.

There were twelve imams in Islam, holy men representing God on Earth. The first imam was named Ali, the son-in-law of Prophet Muhammad. The followers were all male descendants of Muhammad through his daughter Fatimah. Each imam was the son of the previous imam, with the exception of one imam following his brother. The twelfth and last imam is Muhammad al-Mahdi, who disappeared at age 4 around the year 873 and, having no descendants is still considered to be alive as the current imam.

The word Twelver is the description of the largest

branch of the Shia Islam that continues to hold the twelve imams to be the spiritual and political successors to the prophet Muhammad. According to the theology of Twelvers, the Twelve imams are the religious leaders to interpret the Quran, to provide law and justice for the believers. The Sunnis, the other large branch of Islam, argue that the 11th imam did not have a son and this is an historic disagreement that we are going to abstain from taking sides in.

Finally, Buddhism's use of the number twelve is also prominent in the well known twelvefold path. The twelvefold path's elements are the right view, right commitment, right speech, right behavior, right attitude, right diet, right sexual union, right attention, right correction, right breathing and right abiding. The operational word is clearly "right" applied to many facets of our life activity and philosophy.

Most of them are rather self-explanatory, apart from maybe the last, the abiding. That is the practice of behavior leading one to experience the enlightenment, a peaceful state of mind and body. Buddha, the enlightened one, about two and a half millennia ago preached four noble truths and an eightfold path that were later recaptured in the form of the above twelvefold path.

From this chapter we can deduce a deep reverence of humankind toward the number twelve since it was assigned in some capacity to the gods and prophets of all religions. While associated with a religion, the twelve elements of Buddhism are also very practical everyday behavioral advice.

3

Ubiquitous twelve

Twelve became first ubiquitous in our society as a level of categorization. A very early and enduring example is the list of twelve virtues from Aristotle, the Greek philosopher of the 4th century BC. His book on ethics (actually compiled by his son from Aristotle's teachings) contained a chapter on each of the items of the list with detailed explanations.

The virtues listed were: courage, temperance, liberality, generosity, morality, humility, patience, truthfulness, tactfulness, friendliness, modesty, and righteousness. Here the modern English words were used since some of the original Greek words developed a different and sometimes pejorative meaning during the two and a half millennia since Aristotle's time.

For example, the morality was originally magnanimity, that now means something different than just simply the greatness of the soul as Aristotle explained it then. He also made a very clear distinction between liberality and generosity, one being supportive of the downtrodden in a philosophical sense and the other putting their own means to use in such efforts.

Aristotle also categorized the vices into twelve kinds. They were: cowardice, insensibility, meanness, stingi-

ness, immorality, boastfulness, restlessness, insincer-
ity, crudeness, irritability, vanity and maliciousness.
They are almost direct opposites or contradictions of
the virtues again with modern English words carrying
intriguing philosophical interpretations.

As a categorization level, twelve is still in abundant
everyday use. There are the well known twelve steps
of recovery from various addictions, twelve traditions
and even twelve step programs for losing weight. We
are not going to list those, suffice to state our preference for organizing things into lists of twelve.

The word twelve itself is somewhat peculiar and special in most languages. Our English version originates
in the old word Twelf that meant "two left". That is,
two left over the ten. In many languages twelve, along
with eleven, has a name unfitting to the rest of the ten
plus numbers such as thirteen, fourteen, etc.

The prevalence of twelve in systems of measurements
is credited to the English who introduced the length
unit of a foot comprised of 12 thumbs, the origin of
inches. This dates back to the 13th century. Then
came a proliferation of the use of twelve via its multi-
ples into other measurement units. Three feet defined
the yard, or $3 \cdot 12$ inches. Six feet depth is a fathom
(still used by sailors), 660 feet distance ($55 \cdot 12$) is a fur-
long (still used in horseracing) and 5280 feet ($440 \cdot 12$)
became the mile.

On a lighter note, we should mention twelve's ubiquitous presence as a modern drinking unit in the twelve-
pack of various brewed products.

The very important aspect of human life, money, also became influenced by twelve. Historically twelve of the smallest British monetary unit, penny or pence in plural, constituted one shilling. Twenty shillings made a pound. By the same token the original English weight unit pound was comprised of twelve ounces (now 16), but let's stay with the money.

In the 1970's the United Kingdom changed to a metric monetary system and 100 pence became a pound sterling. The image above shows that the preference

for the twelve remained so strong that the winning en-
try of a new pound coin design contest also has twelve
sides. The new, soon to be introduced, coin minimizes
the chance of counterfeiting while still usable in auto-
mated vending and ticket machines.

Other members of the British Commonwealth also
like the concept and the Australian 50 cent coin is
also twelve sided. The replacement of their earlier circular 50 cent coin was for a more mundane reason, the
price of silver. The earlier coin was 80 % silver and
the value of the metal in the coin exceeded its monetary value. Hence the new twelve sided coin is made of
copper and nickel. This coin also has the distinction
of being one of the largest coins in the world at the
diameter of 1.25 inches or 31.5 millimeters. The ob-
verse sides of both the Australian and British twelve
sided coins, of course, feature their common monarch,
Queen Elizabeth.

While the Brits initiated the twelve or dozen based
systems of measurement, it was soon followed by others on the continent. The French in 18th century also
had a foot comprised of twelve inches, that in turn
was of twelve lines and the lines of ultimately twelve
points. A Swedish King, Charles XII has also recom-
mended to use twelve as the basis of measurements.

Money leads to entertainment and the number twelve
is featured in numerous film titles. They range from
the 1957 film Twelve Angry Men to the 1995 film
Twelve Monkeys. The first refers to the twelve members of the jury and the reference to the number twelve
in the second is as cryptic as the whole movie was.

There was also a western Dirty Dozen and the recent drama Twelve Years a Slave.

In sports the professional boxing matches' 12 rounds is notable and in many sports featuring 11 players the fans are called the twelfth player. Some teams even go as far as not letting any player use the number twelve on their jerseys. In bowling twelve strikes are a perfect game and in Canadian football there are twelve players on the field. On the chess board, there is the famous problem of placing eight queens in such a way that they cannot capture each other. The surprising result is that one can do this twelve ways.

There are dozens (a word to become a subject soon) of books and probably hundreds of musical titles referencing twelve. The latter might be because musicians are familiar with the twelve tones that are in an octave. Now this requires a short side track. We have seven notes in an octave, so one could ask why not just halve each of the intervals between them and have 13 tones? Well, the reason is the harmoniousness, or lack thereof.

Splitting into 12 tones the intermediate ones, called half tones, while very close to the seven basic notes, have a different intonation. Human ear considers the integer ratios between sound frequencies pleasing. A full octave difference is a ratio of 2:1. The next integer ratio is 3:2, called the fifth and that is the basis of our 12 pitches or tones.

A teeny bit of arithmetic is required here. Raising $3/2$ to the 12th power results in $1.013 \cdot 2^7$ or almost exactly another power of two. As we said above, our ears do like the ratios of 2:1, hence taking twelve steps of fifths almost repeats the original sound but several octaves higher. Now people with perfect hearing, such as most musicians, can hear the slight difference between the 1.013 and 1.00 times the power of two. The solution is very simply to mistune musical instruments to fifths a bit less than 3/2 ratio to result in the 12th fifth being exactly the higher octave. As crazy as this solution sounds (pardon the pun), it is true and was already proposed by no less than the famous German composer Johann Sebastian Bach almost three centuries ago.

Talking about music leads us to a children's tune. There is an old song, first printed in the 18th century, and widely known in English speaking countries. Its title is "Twelve Days of Christmas" and it has twelve verses. Each day brings a new gift type with the number of the gifts fitting the day. The song is increasingly long as all the gifts of the prior days are repeated.

The first day verse contains only a single gift, a Partridge in a Pear Tree. On the Second day of Christmas the received gifts are Two Turtle Doves (new) and a Partridge in a Pear Tree (from earlier). On the Third day of Christmas the gifts are Three French Hens, Two Turtle Doves and a Partridge in a Pear Tree.

The gifts of the remaining days were in order Four Calling Birds, Five Gold Rings, Six Geese a-Laying, Seven Swans a-Swimming, Eight Maids a-Milking, Nine Ladies Dancing, then Ten Lords a-Leaping, Eleven Pipers Piping and Twelve Drummers Drumming.

Whether the gifts are just repeated on each day or actually repeatedly given is not clear from the song, but in the latter case there is an interesting algebraic computation about the gifts accumulated through the twelve days.

On the first day simply one gift is received. On the second day, since starting the rhyme from start again $1 + 2 = 3$ gifts are received. Adding that to the gift from the first day, the child has 4 gifts after two days. On the third day $1 + 2 + 3 = 6$ new gifts are given and overall 10 gifts are accumulated.

The number of gifts received each day is the sum of the number of gifts received the prior day plus the new gift item of the day. The sequence continues as $10(= 6 + 4), 15, 21, 28, 36, 45, 55, 66$ and the last day produces 78 as the sum of 66 and the twelve drummers drumming.

The total number of gifts accumulated, the sum of all of these numbers is then 364. This is almost one gift for every day of the year. Now if parents would just obey to this rule, all children would be very happy. In a later chapter we will contemplate what would happen if the song was not only for the twelve days but for eternity. But first we will have a bit of arithmetic fun with the number twelve.

4

Dozenal society

Well, whether you, dear reader, consider arithmetic as subject of fun at all, is questionable, but this is a training ground for more serious algebraic things to come.

There are people, called dozenologists, who propose that our arithmetic base should be twelve instead of ten. One of the very first of them, the little known English mathematician Joshua Jordaine published a book titled Duodecimal Arithmetick (sic!) in 1687. He made his case for the introduction of base twelve or duodecimal (from the Latin word of twelve) system for everyday arithmetic besides its use in measurements.

In a base 12, also called dozenal system, we would need different symbols for ten and eleven. The symbol 10 in the dozenal system corresponds to twelve in the decimal system. There were various propositions for the notation for 10 and 11 in the dozenal system. One proposition was made by the French Pascal in the 18th century to use the letters A and B for decimal 10 and 11. We will follow his recommendation.

The question arises: how would one write the numbers above twelve. Decimal numbers 13, 14 and 15 would be dozenal 11, 12 and 13. On the other hand dozenal 1A, 1B, 20 would correspond to decimal 22, 23 and 24. Then 100 in dozenal is 144 in decimal; 1,000 is 1,728; 10,000 is 20,736; and so on.

Jordaine and other advocates of the dozenal system point out the advantages of a dozenal system in representing everyday fractions of importance. For example the fractions

$$1/2, 1/3, 1/4, 1/6, 1/12, 1/24$$

in the decimal system are

$$0.5, 0.33\overline{3}..., 0.25, 0.66\overline{6}..., 0.0833\overline{3}, 0.0416\overline{6}...$$

where some of them contain infinite sequences of repeating digits, denoted by the line over them. On the other hand, the same fractions in the dozenal system are simply

$$0.6, 0.4, 0.3, 0.2, 0.1, 0.05$$

All nice dozenal numbers with single decimal, oops, dozenal, digits where the dot is now the dozenal point. Since no one can argue with the importance of dealing with the above fractions, it seems like dozenal is good thing.

Countering this argument are the cases where the decimal is easy but difficult in dozenal. For example, the decimally simple 0.05, or one twentieth, is an unappealingly infinite number of repeated digits in dozenal as $0.0724\overline{97249}...$. Then there is that troublesome fraction $1/7$ whose representations in both decimal ($0.142857\overline{142857}..$) and dozenal ($0.186\text{\AA}35\overline{186\text{\AA}35}$) require infinite number of digits. It appears that there are pros and cons for dozenal arithmetic.

Before we make a judgment, let us see how we could convert decimal fractions to dozenal. This can be accomplished by the repetitive procedure of multiplying the decimal digits by 12, subtracting the integer part and repeating the procedure. The integer part number then represents the next dozenal digit.

For example, converting decimal $(0.862)_{10}$ to dozenal proceeds as $12 \cdot 0.862 = 10.344$ where the integer part is 10 and in dozenal it is A. This is our first digit. Then $12 \cdot 0.344 = 4.128$ and the second dozenal digit is 4. Doing once more we get $12 \cdot 0.128 = 1.536$, yielding 1. The dozenal result is $(0.A41...)_{12}$ and since we still have a fractional part of the result (0.536), the process would need to be continued. This example number is longer in dozenal digits.

Now we will really stretch the reader's abstraction capability when we apply the reverse of this process to convert a dozenal number to decimal. In this case for a given dozenal number we repeatedly multiply by ten, but in dozenal (by A), and do the same cutting off the integer part. For example, the dozenal number $(0.843)_{12}$ can be converted by the sequence of steps: $0.843 \cdot A = 6.A66, 0.A66 \cdot A = 9.75, 0.75 \cdot A = 6.22, ...$ into the decimal fraction of $(0.696...)_{10}$.

With a calculator we can simply verify both processes as

$$(0.A41...)_{12} = 10/12 + 4/(12^2) + 1/(12^3) =$$

$$0.833\overline{3}.. + 0.0277\overline{7}.. + 0.000578.. \approx (0.862)_{10}$$

and

$$(0.843)_{12} = 8/12 + 4/144 + 3/1728 =$$

$$0.66\bar{6}.. + 0.0277\bar{7} + 0.00173 = (0.6961...)_{10}$$

The appeal is less and less, although in all fairness, the strangeness of the dozenal arithmetic in part is likely just a cognitive dissonance.

Before leaving the dozenal arithmetic, however, let us illustrate the elementary operations in this system for the venturesome reader to follow. For example, here is an addition of two dozenal numbers:

$$38A45B3$$

$$+6108BA5 =$$

$$99B1598.$$

A subtraction example is

$$74B8A6$$

$$-314364 =$$

$$437542.$$

The multiplication requires a bit more thought so we will use simpler numbers:

$$8694 \cdot 24 =$$

$$2A314$$

$$+5618 =$$

$$7B994.$$

Since the long division even in decimal system brings painful memories to most everyday readers, we will refrain from discussing a dozenal division example.

There was also prominent opposition to the duodecimal system, most notably from the famous Lagrange. He pointed out that the abundance of proper divisors of twelve is actually a hindrance rather than an advantage because some of the simple fractions lose their denominator by simplifying and some don't. Instead, he proposed to use a prime number based arithmetic, for example eleven or thirteen. The latter case would have fitted the world had we developed 600 million years earlier and based our number system from celestial observations on that time. Whether we would have developed 13 fingers is another question.

There are still proponents of the dozenal system today. In fact, a so-called Dozenal Society exists both in the United States and in England. Their modern arguments start from the same place that our fascination with twelve was coming from: time. They propose a dozenal clock, that would look exactly the same as today's clock except it would end in ten instead of twelve. An argument for the dozenal clock is based on the fact that the number of minutes in 24 hours is $24 \cdot 60 = 1440$ which is ten times twelve square.

The marking between the (dozenal) hours would also be in units of twelve. Each dozenal hour would have 72 dozenal minutes and each minute 72 dozenal seconds. They would be a bit shorter than in our current decimal system, one minute in dozenal would be equivalent to 50 decimal seconds. On the other hand, 1 second in the dozenal system would be 25/36th of our conventional second.

Using a dozenal clock, a day would be 20 dozenal

hours corresponding to the 24 in decimal. The time of
quarter to midnight in the decimal clock would be the
same, since three twelfth is also a quarter in dozenal.
Similar positions of the clock would mean similar times
measured in different units. Still, the clock below
would require some serious getting used to with a bit
of mental change in measuring time units.

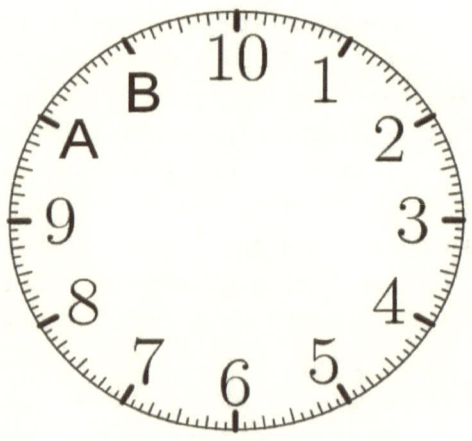

Nevertheless, dozenologists propose to carry the same
division into the realm of geometry. As we see it from
the clock, a full circle is subdivided into 144 equal
parts, but denoted as 100 dozenal parts. One unit of

that is proposed to be called one grad, to distinguish from degree. Hence the 360 degree full circle would be 100 dozenal grads. One grad would be equivalent to one 144th of the full circle, or 2.5 degrees.

This could be advantageous when describing a circle divided into parts. For example, a quarter circle (which we know as 90 degrees) would be 30 grads in dozenal (that is 36 in decimal). Dividing a circle into 2, 3, 4, 6, 8, 9, 12 or even 16 parts can be done without any fractions in dozenal. For example, the latter is 22.5 degrees in decimal, but simply 9 grads in dozenal, a single integer digit.

Dozenologists, who are very passionate about their topic, have some additional arguments in their arsenal. One of them is that the usual keypad arrangement of telephones is already dozenal, well, organized into twelve buttons in four rows of three. The top three rows presenting the nine decimal digits and the bottom row bringing the star, zero and pound signs. In fact, some dozenologists recommend using the star and pound signs for the roles of A, B used above.

It appears that unknowingly, at least as far as the telephone keypad is concerned, we are already sort of a dozenal society.

5

From twelve sides

Let us first consider a simple planar geometric object of twelve equal sides. It is very easy to construct one since we learned how to make a hexagon with a simple compass already in high school. The same compass can be used to halve each central angle of the hexagon, which we know to be 60 degrees, into two halves of 30 degrees and we arrived at the dodecagon with twelve (dodeca) angles (gonos) and sides.

The central angle viewing the sides of the dodecagon is the above constructed 30 degrees since obviously twelve of those adhere to the ancient rule of the complete circle being 360 degrees. The sides of a dodecagon constructed inside a unit radius circle are of length $\sqrt{2 - \sqrt{3}}$, an intriguing combination of factors of the number twelve.

Interesting objects may be generated from polygons by repeatedly extending the sides and intersecting them with each other. Such objects in the case of dodecagon are called the dodecagrams. Depending on whether the intersection with the first, second or third non-neighbor sides is chosen, different dodecagrams may be obtained. The connection of the outermost intersection points still produces a dodecagon, of course.

The image below shows how the extension intersections between every third non-neighbor sides produces an external dodecagon. Furthermore, the extension lines produce an interesting image resembling to a star, hence the process is called stellation from the Latin word for star.

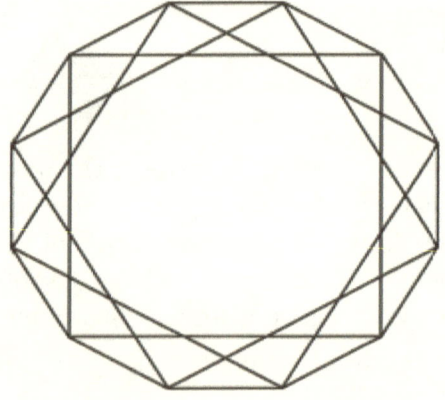

The process was employed already by the ancient geometers mainly for the smaller regular polygons, such as the hexagon and the pentagon, but it is applicable to any regular polygon. Furthermore, the process may be applied to regular polyhedra as we will see it later.

On that note we move onto a three dimensional object of twelve sides, called the dodecahedron. We are now walking on the path of Platonic solids, laid down by Plato around 350 BC. Plato, however, did not complete the path; he only got as far as the dodecahedron shown below.

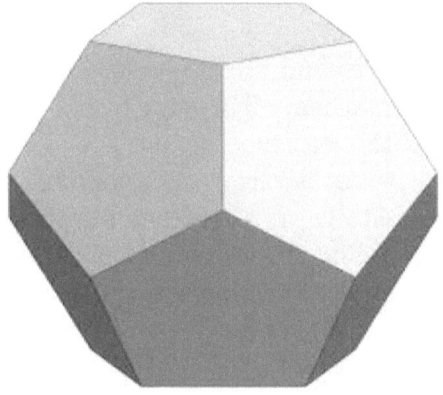

Plato's contemporary, according to some records possibly a friend or a student, Thaeteus added the icosahedron (subject of the next chapter) to the list of regular polyhedra, thereby completing it as there are no more.

Plato nevertheless is now known as the father of this class of solids, mainly due to his philosophical writings about them. He wrote that the dodecahedron was God's model when creating the universe. In his book titled Timaeus he connected the dodecahedron to all aspects of life besides geometry and cosmology, among them physics and biology, subjects of later chapters.

It was Thaeteus who proved that the dodecahedron's twelve sides are regular pentagons. Because of that, the dodecahedron has 20 corners and 30 edges between those corners. The latter number is obtained because each side has 5 edges and 12 sides would give 60, however, each edge is common to two sides, hence only half of them are unique.

Those 30 edges and 20 corners of the dodecahedron gave rise to a mathematical game by the British Hamilton in the 19th century. The subject of the game was to walk the edges of the dodecahedron and reach all corners without ever repeating an edge with the starting and the ending corner being the same. This is, in general, a problem of graph theory. Graphs are comprised of straight edges between a set of points usually in a plane.

Hamilton knew in advance, as any self-respecting mathematician would also do before posing a challenge, that the problem can be solved. Such a path in general graph theory is now called a Hamiltonian cycle.

Mathematicians also like coloring and answer questions like how many different colors would be needed to cover the sides of a polyhedron such that no neighboring sides are of the same color. Or, how many colors are needed to color all edges without two adjacent edges sharing the same color. Finally, the question of how many colors would be needed to color the corners in a way that no two adjacent corners share the same color. The answers to these questions for the dodecahedron are four for the sides, three for both the edges and the corners.

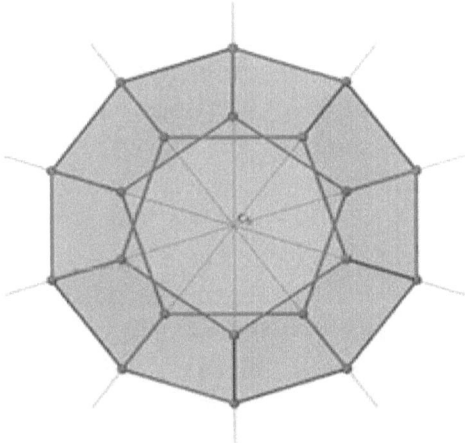

   The visual image of the dodecahedron shows spectacular symmetry with many axes as shown above.

There are 60 ways to rotate the object and still retain its positional appearance. First, a single rotation
is possible by 180 degrees around the axis drawn between the 15 pairs of the center points of the opposite
edges. One can also rotate around an axis drawn between 10 pairs of opposite corners by 120 degrees 2
different times. Finally, rotation is possible around an
axis drawn between the 6 pairs of the center points of
opposite sides by 72 degrees 4 different times. Adding
to that the rotation around any axis by zero degree,
the so-called identity operation that obviously retains
the original orientation, the sum of $15\cdot1+10\cdot2+6\cdot4+1$
produces the 60 symmetries.

Besides the appeal arising from symmetry, there must
be some source of it in proportions. Much has been
written about the so-called golden ratio, a proportion
of geometric components extremely pleasing to the human eye. The algebraic value of that ratio is easily
computed and well known to be $(\sqrt{5}+1)/2$ and denoted commonly by ϕ. Well, maybe not surprisingly,
the length of the edge of a dodecahedron inscribed into
a unit sphere is $2/(\phi\sqrt{3})$, related to the golden ratio.

Further proof of the appeal is notable in the coordinates of the corners of the dodecahedron placed at
the origin of the x, y, z coordinate system. For algebraic simplicity and to make the point more demon-
strative, let us assume that the dodecahedron is inscribed into a sphere of radius $\sqrt{3}$. Then the coordinates of the twenty corners are at (x, y, z) val-
ues of $(\pm1, \pm1, \pm1)$, $(0, \pm1/\phi, \pm\phi)$, $(\pm1/\phi, \pm\phi, 0)$ and
$(\pm\phi, 0, \pm1/\phi)$, an amazing collection of golden ratios.

This phenomenal gathering of golden ratios continues inside the dodecahedron. It has been well known since Plato's time that a cube (regular hexahedron) can be fitted inside a dodecahedron five different ways. One such nesting of a cube is shown below.

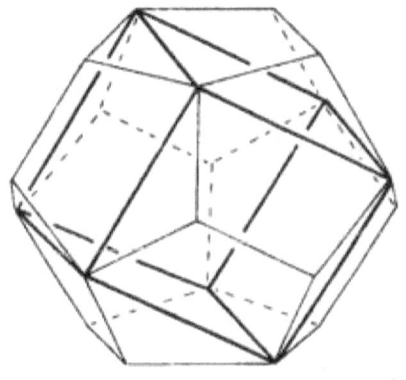

This helped to overcome the conceptual conflict between our visibly three dimensional (e.g. cubic) world and Plato's suggested dodecahedral universe. One can interpret this as life in the three dimensional world nested into the dodecahedral universe.

Let us now investigate the relationship of this cube with its circumscribed dodecahedron. The ratio of the edge of the dodecahedron vs. the inscribed cube's edge is $1 : \phi$. The ratio of the volume of the dodecahedron vs. the cube is $1 : 2/(2 + \phi)$. The golden ratio all over again. It is said that the dodecahedron has a golden frame!

The dodecahedron's spectacular symmetry and its golden proportions made it a popular component of generalizations. Perhaps furthering Plato's thinking, the French mathematician Poincare in the late 19th century, being conscious of the infinity of space, proposed a dodecahedral universe where the opposite sides of the dodecahedron were twisted and connected. This genial construction made an infinite but bounded space, sort of a contradiction, of course. But all bets are off when one is dealing with infinity, as we will see very profoundly later.

As far as Poincare's dodecahedral universe is concerned, however, it is actually being taken seriously in cosmologist circles and there are attempts to prove or disprove it. The connections via the opposite sides are not geometrical but rather the so-called wormholes of space time. The volume of the dodecahedral universe would be 2.79 times the cube of the radius in comparison to the spherical universe's 4.18 times the cube of the radius. We apparently have less space in a dodecahedral universe.

Finally, one can create another polyhedron from the regular dodecahedron. Connecting the centers of the dodecahedron's twelve sides (denoted by the dots in the image below) we obtain the edges of an internal polyhedron.

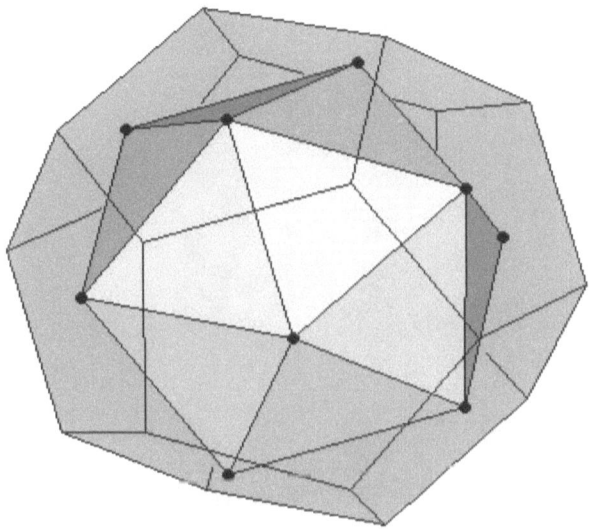

This polyhedron is, of course, Thaeteus' icosahedron, the subject of our next chapter, with twelve corners and twenty sides, the reverse of dodecahedron's numbers, but also having 30 edges.

6

To twelve corners

No, we are not talking about the corners of the Earth yet, although there will be also twelve of them, at least at the beginning. As we saw at the end of the last chapter, we have a new solid with twelve corners, called the icosahedron shown below.

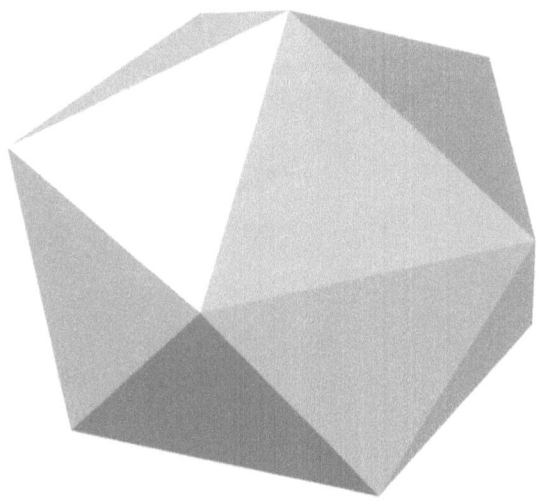

Because of the relationship between the dodecahedron and icosahedron, they are mathematically called the duals of each other. In order to make some of the later comments easier to interpret, let us visualize a see-through image of the icosahedron as shown below.

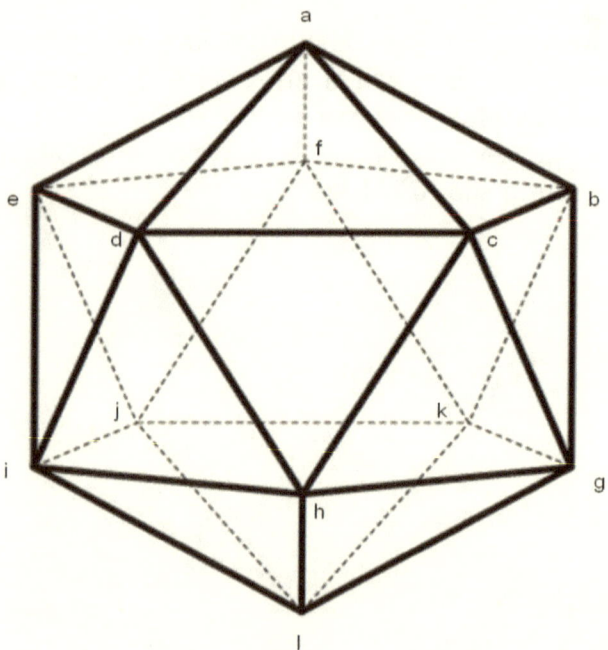

While the sides are now regular equilateral triangles, regular pentagons are also lurking inside. The two horizontal ones are immediately noticeable, connecting corners (b, c, d, e, f) and (g, h, i, j, k), but the others require a bit of looking. These pentagons will be of high importance in a later discussion.

We will follow the topics discovered in the last chapter here as well, starting with the stellation method we have seen in the plane with the dodecagon, that is also possible in space.

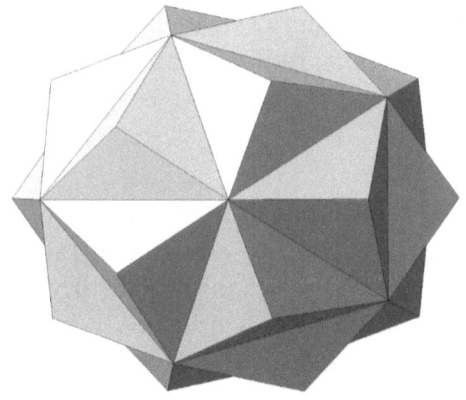

Starting from an icosahedron, the stellation process by extending and intersecting sides results in intriguing images. The one above is one of the first 59 possible stellations of the icosahedron, generated by using the software program Great Stella: Polyhedron Navigator by Robert Webb, available at www.software3d.com.

The problem of Hamilton is also solvable on the icosahedron: there exists a circular edge walk that visits all corners without repeating any edge. The coloring of the icosahedron is a bit more difficult; one needs 4 colors to color the corners, 5 colors to color the edges and 5 colors to color the sides, with the conditions established in the last chapter.

The external proportions of the icosahedron are also spectacularly golden. The length of the edge of a regular icosahedron inscribed into a sphere of radius r is $l = 2r/\sqrt{\phi^2 + 1}$, where ϕ is again the golden ratio. The length is about $1.05146 \cdot r$, or the edge of an icosahedron with any radius is about 5% more than the radius. This will be a ratio reused in a later chapter.

The (x, y, z) coordinates of the corners of an icosahedron are again related to the golden ratio. For algebraic simplicity, we select an icosahedron with edge length of 2 which corresponds to $r = \sqrt{\phi^2 + 1}$. Then the twelve corners are at the locations $(0, \pm 1, \pm\phi)$, $(\pm 1, \pm\phi, 0)$ and $(\pm\phi, 0, \pm 1)$. Note we have only 3 quadruples (due to the $\pm$ options) totaling twelve.

The surface of the icosahedron is a topic of interest and it is easily computed as twenty times the area of a side. Each side is an equilateral triangle with the above computed edge length, hence the total surface area is $5\sqrt{3}l$. Finally the volume is comprised of twenty tetrahedra over the sides with apex at the origin of the coordinate system and yields $5\phi^2 l^3/6$. This is about 60 % of the volume of the circumscribed sphere, so there is room in it, a fact we will also use later.

This leads us to the internal proportions of the icosahedron that are similarly golden. This is best demonstrated by the three internal rectangles shown below. They are so-called golden rectangles because the ratio of their sides is the golden ratio.

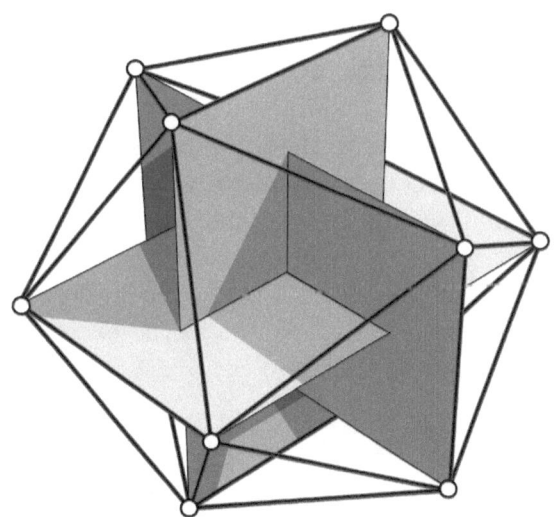

Then there are golden distances as well. The distance from the center point of the volume to the middle of a side, for example, to the middle of pentagon (b, c, d, e, f) is $l\phi^2/(2\sqrt{3})$. The distance from the center to the middle of an edge, for example the midpoint between (a,b), is $l\phi/2$ and the distance from the center to a corner is $l\sqrt{\phi^2 + 1}/2$. Golden ratio in all!

Further exploring the earlier see-through image of the icosahedron, another interesting distance appears. The distance from any corner (for example (a)) to the center of a pentagon created by the five neighbors of that corner (b, c, d, e,f) is $l/\sqrt{\phi^2 + 1}$. Similarly, the distance between any pair of parallel embedded pentagons is $l\phi/\sqrt{\phi^2 + 1}$. An example of that is the distance between the centers of the (b, c, d, e, f) and (g, h, i, j, k) pentagons.

That many golden ratio numbers bound to raise some suspicion. In order to qualm the fears of something being wrong, let us evaluate the formula for the distance of a corner to the center numerically. On a calculator using 5 decimal digits we get $0.95105 \cdot l$. We established earlier that the edge length is $l = 1.05146 \cdot r$. Hence the numerical distance from the corner to the center is $0.99999... \cdot r$, that would be the exact radius in higher precision as it should be.

The icosahedron will have the same number of rotational symmetries as the dodecahedron and the definition of the axes of symmetry will not change either. But the number of symmetries regarding a particular type of symmetry axis will change.

Axes of the first type also pass through the midpoints of two opposite edges and there are still 15 of them since the icosahedron also has 30 edges. These may be considered as single rotations of 180 degrees, resulting in 15 rotational symmetries.

Axes of the second type also pass through two opposite corners, however, there are 6 of such pairs from the icosahedron's 12 corners. These may be considered as 4 rotations of 72 degrees, resulting in 24 rotational symmetries.

Axes of the third type also pass through the centers of two opposite sides, but there are 10 of them due to the icosahedron's 20 sides. These may be considered as two rotations of 120 degrees, resulting in 20 rotational symmetries.

A first type of axis, for example, is the line between the midpoints of the (a,b) and (i,l) sections. An example of the third type of axes is the line between the center points of the (a,b,c) and (i,j,l) triangles. All of the second type of rotational axes are easily described in terms of our image as the lines between opposing pairs of corners (a,l), (b,i), (g,e), (f,h), (j,c) and (k,d).

For example, the clockwise rotation by 72 degrees about the axis spanned by the selected top (a) and the bottom (l) corners results in the ordering of a, c, d, e, f, b, h, i, j, k, g, l. As the axis of rotation was between a and l, they have not been moved, however, the other ten corners have rotated. Further rotations by 144, 216, 288 degrees will provide the other 3 symmetry positions regarding to this particular axis type. Since there were 6 pairs of such, these produce 24 different rotational symmetries.

Similar considerations about the other axis types

produce 15 and 20 other symmetries. Finally adding the identity symmetry operation of rotating by zero degree about any of the axes results in the total number of rotational symmetries of the icosahedron to be 15+24+20+1=60, same as that of the dodecahedron.

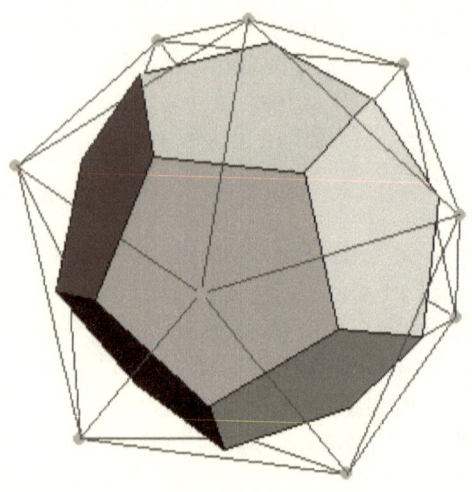

Earlier we created an icosahedron from the dodecahedron, we now do the reverse: create a dodecahedron from an icosahedron, as shown above. This could be done by connecting the centers of the icosahedron sides to centers of neighboring sides sharing an edge with lines that become the edges of the dodecahedron.

This intriguing process, going back and forth between dodecahedron and icosahedron, also illustrates their dual nature manifested in their shared number of symmetries. This might also be the possible reason for nature's preference for objects of dodecahedral, and especially icosahedral shapes, as we will explore it in the penultimate chapter.

The rotational symmetry group of the icosahedron described in terms of the corner letters present a certain set of permutations of twelve letters. The total number of possible permutations of twelve letters (or objects) is, however, a much larger number. That includes permutations that cannot be described in terms of rotational operations of the icosahedron.

Some of those permutations can be reached by translational symmetry operations on the icosahedron. Such symmetries are the reflections through planes going through the center of the solid, like the planes of the golden rectangles shown earlier. And, as a matter of fact they also count 60. Overall the icosahedron has 120 distinct symmetries.

This total number, however, is still far lower than the possible number of permutations of twelve objects. Those permutations carry us into the land of algebra.

7

Symmetry of twelve

Twelve has many spectacular geometrical symmetries as we already saw in the past two chapters in the dodecahedron's twelve sides and the icosahedron's twelve corners. There is, however, a less visible algebraic manifestation of the symmetry of twelve. To see that we need to venture into an area of mathematics called group theory.

To keep the discussion at the intended level of the book, we will define things verbally as much as possible. So, here we go: A mathematical group is a collection of operations on some object or objects. Any two operations of that collection executed successively must result in another operation that is also part of the same collection, i.e. the group.

The best way to conceptualize such groups is still to visualize them using a geometrical object. For example, considering a simple equilateral triangle, rotating it by 120, 240 or 360 degrees, the positional appearance of the object will not change. Furthermore executing two operations successively, for example rotations by 120 and 240 degrees, will result in another operation of the group, 360 degrees. We will consider the operations circular, 480 degrees rotation is identical to 120 degrees, etc.

This example group is also of a specific type, called a cyclic group. A cyclic group is comprised of operations that could all be reached by the successive execution of a single operation of the group. In this case the rotation by 120 degrees is the single operation as shown in the figure below.

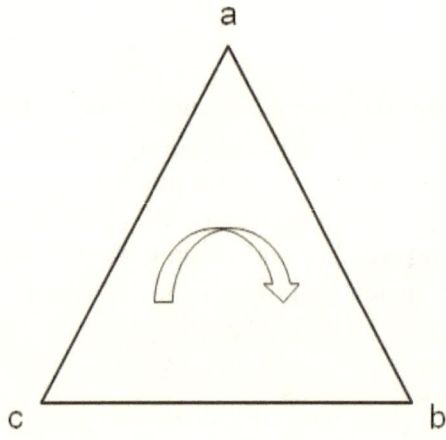

The number of times the operation is repeated until the original orientation is returned is the size of the group. In the case of our example it is 3 since rotating the triangle by 120 degrees 3 times will return it to its original orientation. This could be verified by following the 3 corners of the triangle denoted by a, b and c, and rotate until the same orientation is achieved.

If the size of a group is a prime number, as it is in our case, then the group is called a prime cyclic group. Such groups are very special because for every prime number there is only one cyclic group of that size. To describe groups of higher size the algebraic operation of permutations is commonly used. Permutations are different arrangements of a set of algebraic objects in various ways resulting in a group. This does not mean that such groups do not have a possible geometric visualization. In fact, we will visualize a spectacular group in connection with the permutations of our number twelve via the icosahedron shortly.

The number of permutations of 3 objects is easily interpreted by using our triangle. The three corners (a, b, c) may also be in the sequence of (c, a, b) and (b, c, a). These are simply the 3 rotations we considered above. They may, however, also be in the order of (a, c, b), (b, a, c) and (c, b, a). These arrangements cannot be reached by rotating the triangle, but can be reached by reflecting the triangle through a line from a corner to the center of the opposite side. The (a, c, b) arrangement is the mirror image of the (a, b, c) arrangement by reflecting through the line from the top corner (a) to the middle of the (b,c) side.

Another way to look at this scenario is by considering a single operation of swapping two members. In this view reaching (a, c, b) from (a, b, c) is a single swap of (b) and (c). On the other hand, to reach (b, c, a) from (a, b, c) requires two swaps: first (a) with (b) followed by (b) with (c). Note that the first three arrangements obtained by rotations in the geometrical sense require two, hence even number of, swaps. Furthermore, it is notable that if a permutation can be reached by an even number of operations (e.g. swaps) then it cannot be reached by an odd number of operations. The reverse of this statement is also true and it is true for any order permutation groups.

The total number of permutations for our three object example of the triangle is 6, the symmetry group of the three objects, commonly denoted by S_3. Those first three arrangements, that could be reached by rotations, are called the rotation symmetry group. Since these were the even number of swaps, the total number of even permutations for this group is 3, denoted by A_3. The notation comes from S for symmetry and A for alternating.

A_3 is a subgroup of the S_3 symmetry group and cannot be further decomposed into other simpler groups, hence it is called a simple group. The important consequence of this is, stated in general terms now, that the group of all even permutations of a certain number of objects form a simple group.

Following the logic, we can see that for four objects $S_4 = 24$ and $A_4 = 12$. This case enables us to define so-called multiple transitive groups that move a certain number of objects into another set of the same number of objects. For example, a group that sends any 2 objects to any other 2 objects of the members is called 2-fold transitive. Consider a square whose

corners are named (a, b, c, d). A 180 degree rotation
of our square will send (a, b) to (c, d). However, it is
visible on the image below that there are no symmetry
operations (neither rotations nor swaps) that will send
(a, b) to (b, d), hence the group of permutations of 4
objects is only 1-transitive.

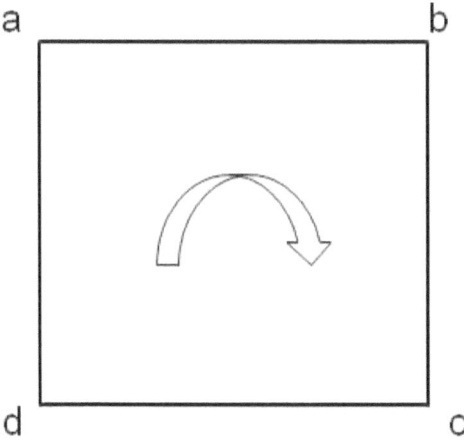

The total number of permutations of 12 objects, S_{12}
is staggeringly large: 479,001,600. This number is just
shy of half a billion. This, in the terms of the icosahedron, is the number of ways we can label the corners
of the icosahedron. The number may be derived by

simple logic: there are 12 ways to pick the first corner, 11 to pick the second, 10 for the third, and so
on until there is only one choice for the last. Hence
$12 \cdot 11 \cdot 10... \cdot 3 \cdot 2 \cdot 1 = 479,001,600$.

Half of this number is the number of even permutations, A_{12}, still a rather large 239,500,800. Since
we are talking about a group of twelve, it is natural
to attempt to visualize this in terms of an icosahedron. The contemporary British mathematician John
Conway proved that all members of that group can
be generated by a series of 72 degree rotations of an
icosahedron.

Using the sea-through icosahedron with alphabetical
corners from a prior chapter and visualizing via one of
the pentagons inside, the counter-clockwise rotation
around the axis (a, 1) moves the corners of the pentagon (and permutes the members of the group) from

$$f$$
$$e \qquad b$$
$$d \quad c$$

into

$$b$$
$$f \qquad c$$
$$e \quad d$$

There are, however, some gems of symmetric simple
groups hidden in the set of twelve objects that are of
smaller size than M_{12}. Such are called sporadic simple
groups and first found by the French mathematician

Emile Mathieu in the 1850's. Remarkably, the very first one he found was related to our number twelve and is called M_{12} in his honor.

The size of that group is only 95,040, certainly a much smaller number by about 4 orders of magnitude than S_{12}. Furthermore, this group Mathieu found in permutations of 12 objects is 5-fold transitive. There is an element in M_{12} that moves any set of 5 elements to any other set of 5 elements, a true 5-fold transitive group.

The size of this group is the same as the number of possible ways one can select 5 members out of a group of 12. This can be achieved by the following: there are twelve ways to select the first, 11 ways to select the second and so on. By the time the fourth member is selected there are 8 more left, so there is 8 ways to select the last. This produces $12 \cdot 11 \cdot 10 \cdot 9 \cdot 8 = 95,040$, which is the size of Mathieu's group.

Conway also proved that an operation of rotating 5 selected corners clockwise and at the same time rotating 5 other corners counterclockwise will produce a member of the M_{12} group.

Let us view this operation with the names of the corners used above. Executing the operation of clockwise rotation by 72 degree, the corners of the lower pentagon move from

$$j \ k$$
$$i \qquad g$$
$$h$$

into

$$i \ j$$
$$h \quad k$$
$$g$$

The simultaneously executed counter-clockwise rotation by 72 degrees was shown just above.

These rotations can actually be demonstrated in practice by placing a ball into the center of the icosahedron and have twelve balls touch it. This is based on the so-called geometric kissing problem in three dimensions, that is, finding the number of spheres that can touch a central sphere. The solution to this problem is, of course, twelve.

This was an interesting contest in the 18th century between two English scientists, Newton, who suspected twelve but could not prove it, and Gregory, who proposed 13, which turned out to be incorrect. The actual radius of the spheres can be computed from the golden distances in the icosahedron presented earlier.

Mathieu's group is the smallest sporadic simple group of 5-fold transitivity that exists. He himself found its bigger brother now called M_{24}, which is also a group of 5-fold transitivity. It represents the smallest simple sporadic group of permutations of 24 objects and its size is formidable, 244,823,040, but still much smaller than S_{24} or even A_{24}. This size is again computable by the possible ways of selecting 5 members out of the group of 24, which is $24 \cdot 23 \cdot 22 \cdot 21 \cdot 20$.

The family of sporadic groups is rather exclusive. There are only 26 such groups and Mathieu found 5 of them, including the smallest of the sporadic simple groups, now called M_{11} of size 7,920 with 4-fold transitivity. Due to the selection of 4 from 11 objects, the size comes from computing $11 \cdot 10 \cdot 9 \cdot 8$. He also found two additional groups between the above twelve-related members, the groups M_{22} and M_{23}.

Conway also found 3 sporadic groups, named Conway's constellation, and they are really large. But the two largest sporadic groups, affectionately called by mathematicians the Baby Monster and the Monster (also named by Conway), have 34 and 54 decimal digits, respectively.

The Monster is a group of permutations of 196,883 objects and also is the symmetry group of 26 dimensional strings of the string theory. This connection is called the Monstrous Moonshine conjecture.

But the two Mathieu groups M_{12} and M_{24} are the only 5-transitive sporadic groups. No more such gems exist. Somehow the numbers twelve and five are intrinsically intertwined in algebra as well as in geometry where we saw before: the 12 sided the dodecahedron having 5 corners, or the 12 corners of the icosahedron having 5 connecting sides. This relationship is a fundamental component of the strength of the symmetry of twelve.

8

From a sublime number

When talking about algebraic properties of numbers, they can be characterized by the specific classes to which they belong to. But first let us see what classes twelve does not belong to. It is not a prime number, not a square number, and not a Fibonacci number. Despite not belonging to these prestigious number classes, twelve has still very intriguing algebraic characteristics related to other classes and it will belong to an absolutely exclusive class giving the title of this chapter.

Since twelve is not a prime number, it is a composite number. It is in fact a highly composite number with a large number of divisors exceeding all of its predecessors. The number 12 may be divided by 1, 2, 3, 4 and 6, in all 5 divisors. The number with the highest number of divisors lower than 12 is 6, with 3 divisors. The next highly composite number is 60 with a lot of divisors (1, 2, 3, 4, 5, 6, 10, 12, 15, 20, 30). Incidentally all highly composite numbers have 12 as one of their divisors.

Twelve is also a so-called abundant number. In fact, it is the very first of such numbers. Abundant numbers are those, exceeded by the sum of their divisors. For 12 that is $1 + 2 + 3 + 4 + 6 = 16$, which is 4 greater (the amount of abundance) than 12 itself.

The divisors of twelve when we include itself elevates it into a really exclusive class in which there is only one another member. The class of sublime numbers consists of numbers whose number of positive divisors (including the number) is a perfect number and whose divisors also add up to another perfect number. Perfect numbers are those for which the sum of all proper divisors (excluding the number) add up to the number.

Let us see that for twelve. It has 6 positive divisors (1, 2, 3, 4, 6, 12) and the sum of those divisors (1+2+3+4+6+12) equals 28. Now 6 is perfect because its proper divisors (1, 2, 3) add up to itself. Similarly, $1+2+4+7+14 = 28$ makes it a perfect number. Mind you, twelve itself is not a perfect number. The reason for the exclusivity is that at this moment the only other sublime number known is the $2^{126}(2^{61} - 1)(2^{31} - 1)(2^{19} - 1)(2^7 - 1)(2^5 - 1)(2^3 - 1)$, a number that has 76 decimal digits. That is really very large since the total number of particles in the known universe is estimated to be of 80 digits long.

Then we find joy in the number twelve as it is a Harshad ("Joy giver" in Sanskrit) number. Such numbers are divisible by the sum of their digits in a particular base. Clearly 12 is divisible by $1+2 = 3$ in the decimal system. There are many such numbers in the decimal system but not many of them are Harshad numbers in other bases. Incidentally, 12 is a Harshad number in all bases except for base 8. Maybe choosing 12 as the base of our number system, as proposed by the dozenologists a few chapters prior, has some merits after all since the dozenal 10 is divisible by $1+0=1$.

Another class twelve belongs to is the class of pan-arithmetic numbers. Such are positive integer numbers below which all other positive integers can be obtained as the sum of the divisors of the number. The divisors of 12 are $1, 2, 3, 4, 6$ and all the numbers below 12 can be expressed with their sums. For example 10 is $1 + 3 + 6$, and so on for the others.

Moving onto geometric relations, twelve is also a so-called rectangular number as a product of two consecutive integers (3 and 4). But more interestingly it is also a polygonal number. Specifically and not surprisingly, 12 is a dodecagonal number.

The numbers are easy to generate algebraically as members of the sequence $5n^2 - 4n$ where n is the sequence number, 2, 3 and so on. Clearly for $n = 2, 3$ we obtain $12, 33$. The dodecagonal number sequence is not a very rapidly increasing sequence, the next few members are $64, 105, 156, 217$ and 288. From this it is noticeable that they are alternating between even and odd numbers, that fact generally true for all polygonal numbers.

Such numbers also represent a particular polygon in their arrangement and are generated by counting the node points of a progression of polygons. The progression is created by successively extending the sides of the polygon once more in each stage generating additional interior node points. Note, when counting around the circumference every side brings two nodes: on at the corner and one in the middle.

The geometric process would start with a single dodecagon, hence 12 is the first dodecagonal number due to the 12 corners of the dodecagon. The first two dodecahedrons generated by this process and the partial numbering is shown in the figure below.

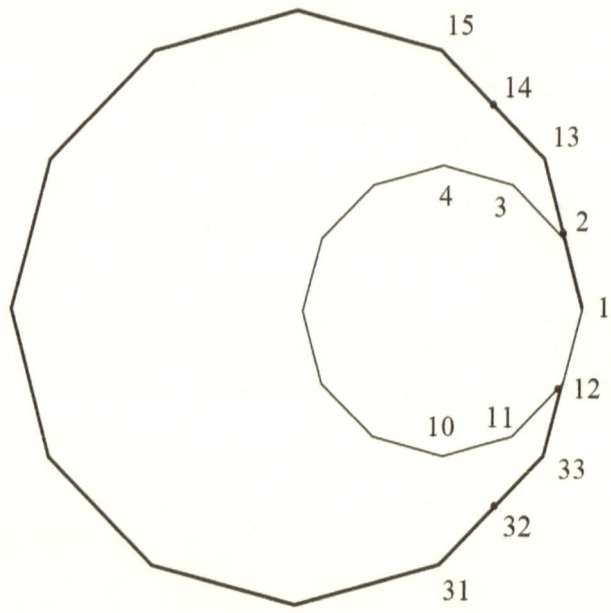

The second dodecagon with a midside node in each edge brings 24 new nodes. However, three of those are shared with the first, hence the next dodecahedral number is 33. This comes from the original dodecagon's 12 plus the second's 12 corner points plus the 12 newly created node points less the three common corners with the first dodecahedron.

Talking about general patterns related to twelve, Fermat, the French amateur mathematician (a lawyer by training and a member of the French parliament) posited that every number is the sum of at most 12 polygonal numbers. He claimed that he proved it, but his proof was never found. The general proof, that eluded even the famous Gauss, was finally found almost two centuries later by Cauchy at the beginning of the 19th century.

One may consider 1 also to be a dodecagonal number as it algebraically satisfies the generating formula: $n = 1$ results in $5n^2 - 4n = 1$. It is, however, hard to see the relationship with the dodecagon, and in fact 1 does satisfy the generating equation of any polygonal number. For example, the generating formula for the hexagonal numbers is $2n^2 - n$ which does give 1, 6 and 15 for $n = 1, 2, 3$. The strict mathematical list of any polygonal number starts with 1, so 1 is the ultimate polygonal number, it is an n-gonal number.

Some of the dodecagonal numbers are perfect squares. Using the complete list of indices starting with 1, the fourth dodecagonal number $D_4 = 64 = 8^2$, the 25th dodecagonal number is $D_{25} = 3025 = 55^2$. Both indices are square numbers by themselves. This rule seems to hold. The dodecagonal numbers that are perfect squares always have an index that is also a perfect square. For example $D_{13^2} = D_{169} = 142129 = 377^2$ and even $D_{34^2} = D_{1156} = 6677056 = 2584^2$.

There are other interesting appearances of the number 12 in deeper mathematical areas. Such an area is

lattice theory. This theory deals with polygons whose
corner points are located in the lattice of the integer
coordinates of the x-y plane. A subset of such polygons
are convex when the line connecting any two corners
of the lattice is inside the polygon in its entirety. We
can further restrict our attention to polygons whose
interior consists only of the origin of the coordinate
system, such as shown below.

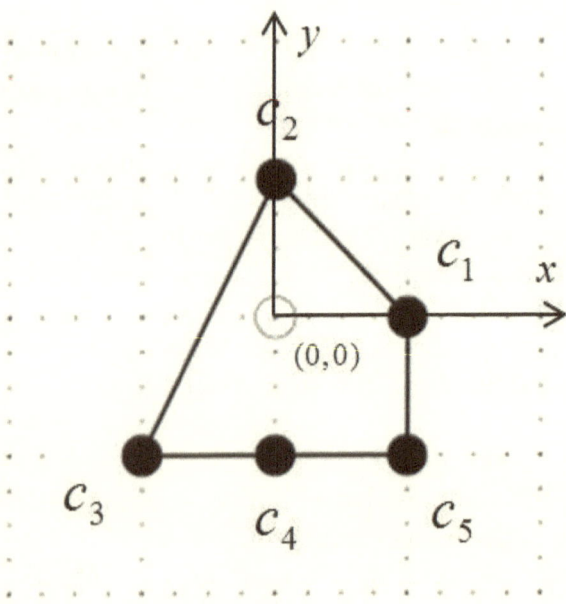

For such polygons a dual polygon is defined as follows. Assume the corners of the polygon denoted in
a counter-clockwise order as $c_1, c_2, ..., c_m$ where m is

the number of corners of the polygon. The corners
of the dual polygon then are created by computing
the coordinate differences of two points, respectively.
For example, $d_1(x, y) = c_2(x, y) - c_1(x, y)$. In the example shown $c_1(x, y) = (1, 0)$, $c_2(x, y) = (0, 1)$ hence
$d_1(x, y) = (-1, 1)$.

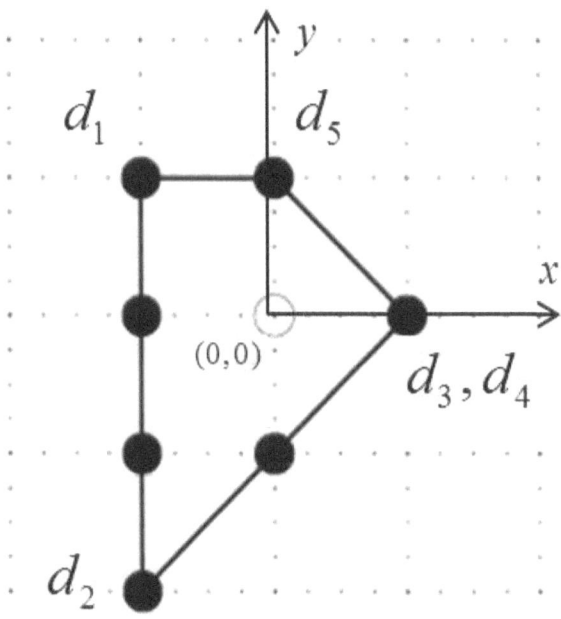

Successively applying the same rule, all corners of
the dual polygon $d_1, d_2, ..., d_m$ may be found. Some
of them may be duplicates, for example, $d_3(x, y) =
d_4(x, y)$. The reader can verify this from the coordi-
nates shown in the figure. We also add to the dual

polygon as corner points those lattice points that are crossed by a side of the dual polygon. The interesting conclusion is that the sum of the number of corner points of the primary lattice polygon and the dual lattice polygon is always twelve. One can verify this with our example as $5 + 7 = 12$. The proof points to some elaborate areas of algebraic geometry that we will not follow, but twelve is involved again.

There are peculiar ratios resulting in 12 between numbers containing all nine digits only once:

$$45792/3816 = 12$$
$$73548/6129 = 12$$
$$89532/7461 = 12$$
$$91584/7632 = 12$$

One can make the case for this being a simple consequence of permuting the nine digits in a certain order or acknowledge a remarkable demonstration of another specific role of twelve in our algebraic system.

The befuddling appearances do not end there. The well known last theorem of Fermat states that equations of the type $a^n + b^n = c^n$ have no solution for any number other than $n = 2$. This theorem occupied the minds and lives of many mathematicians for hundreds of years. Finally, a decade or so ago, it was proven to be true. However, when scientists were trying to find counter examples by exhaustive computer searches, several peculiar "near solutions" appeared.

You may have guessed it right: they were related to our number twelve. It was found that $1782^{12} + 1841^{12}$

is almost equal to 1922^{12} and $3987^{12} + 4365^{12}$ is almost equal to 4472^{12}. The error in the latter case is only 0.000000002%. That is very close, indeed. The exact numerical relation is

$$3987^{12} + 4365^{12} = 4472.00000000070576171875^{12}.$$

The fact that the exponent of 12 comes that close to produce counter examples to Fermat's theorem just adds to the algebraic mystique of our number.

While we are at expressions containing exponents, we must mention that the 5th power of twelve is the smallest such number that can be expressed as a sum of 6 other 5th powers. Specifically,

$$4^5 + 5^5 + 6^5 + 7^5 + 9^5 + 11^5 = 12^5$$

is another intriguing expression with the divisors of twelve and its lesser primes involved.

Furthermore, the mild looking relation of $1225 = 35^2$ can be extended by repeatedly inserting 12 between the 1 and 2 digits, which results in the sequence of

$$112225 = 335^2$$

$$11122225 = 3335^2$$

$$1111222225 = 33335^2$$

$$111112222225 = 333335^2$$

and so on until infinity, a subject that also has some relationship with twelve.

9

To a sum of infinity

Let us recall the Twelve Days of Christmas song with the ever increasing sequence of gifts on each day. The cumulative amount of gifts received during that twelve day period was interestingly 364. But, why do we stop at receiving gifts at the twelfth day? Why don't we continue the process forever?

That is a conceptually simple problem of adding up the integer numbers ad infinitum. One would guess that the result is an infinite number of gifts. Well, Leonhard Euler, the famous Swiss mathematician of the 18th century produced a tantalizing occurrence of our number twelve in that sum

$$S = 1 + 2 + 3 + 4 + \ldots = -\frac{1}{12}.$$

This bizarre result is worthy of a deeper evaluation but first we will attempt to prove it with elementary algebraic tools.

For this we create a simpler infinite sequence made of positive and negative ones:

$$S_1 = 1 - 1 + 1 - 1 + 1 - 1 + \ldots$$

Then we subtract it from one as follows

$$1 - S_1 = 1 - (1 - 1 + 1 - 1 + 1 - 1 \ldots) =$$

$$1 - 1 + 1 - 1 + 1 - \dots = S_1.$$

Hence $1 - S_1 = S_1$ from which elementary algebra gives the solution of $S_1 = \frac{1}{2}$. Note that we did not close the parenthesis due to infinity being undefined.

We also create another alternating sequence as

$$S_2 = 1 - 2 + 3 - 4 + \dots.$$

We now add this sequence to itself but with a specific spacing by introducing a zero as:

$$S_2 + S_2 =$$
$$1 - 2 + 3 - 4 + 5 - \dots +$$
$$0 + 1 - 2 + 3 - 4 + \dots =$$
$$1 - 1 + 1 - 1 + 1 - \dots = S_1.$$

What we see is that $2 \cdot S_2 = S_1 = \frac{1}{2}$ from which we conclude that $S_2 = \frac{1}{4}$.

Let us now subtract this from our original sequence with appropriate spacing again as

$$S - S_2 =$$
$$1 + 2 + 3 + 4 + 5 + \dots.$$
$$-(1 - 2 + 3 - 4 + 5 - \dots.) =$$
$$0 + 4 + 0 + 8 + 0 + 12 + \dots. =$$
$$4(1 + 2 + 3 + 4 + \dots. = 4 \cdot S.$$

Since we found above that $S_2 = \frac{1}{4}$ we have $S - \frac{1}{4} = 4S$ from which it follows that $S = -\frac{1}{12}$. It appears we proved this unfathomable result.

The reader with a suspicious mind might ask whether this is one of those algebraic tricks resulting in implausible results because of a sleight of hand not noticeable in the process. In our case it could be due to the fact that, while we included the infinite sequences in the arithmetic, we never actually computed their sums.

To alleviate this fear we now present the topic the way Euler proved it. Readers with less interest and foundation in algebra could skip to the end of this chapter to avoid some mental strain, although the presentation is as simplified as possible without losing track of the process.

Euler invented a very specific function, the zeta function, with infinite number of terms

$$\zeta(s) = \frac{1}{1^s} + \frac{1}{2^s} + \frac{1}{3^s} + ...$$

Here the s is in general any number. Recalling from high school that $n^{-1} = \frac{1}{n}$ for any number, the zeta function at $s = -1$ has the value of

$$\zeta(-1) = 1 + 2 + 3 + 4 + 5 + ...$$

which is the sum that we are trying to compute. Euler then computed the expression

$$(1 - \frac{2}{2^s})\zeta(s) = \frac{1}{1^s} - \frac{1}{2^s} + \frac{1}{3^s} - \frac{1}{4^s} + ...$$

Substituting $s = -1$ into this again the result is

$$(1 - 4) \cdot \zeta(-1) = 1 - 2 + 3 - 4 + 5 - ....$$

The right hand side's sum was shown above from the tricky, but simple addition process to be $\frac{1}{4}$, hence

$$-3\zeta(-1) = \frac{1}{4}$$

Dividing both sides by -3 and substituting $\zeta(-1)$ from above the proof of our original problem emerges

$$1 + 2 + 3 + 4 + 5 + \ldots = -\frac{1}{12}$$

Twelve appears in other values of the zeta function. For example,

$$\zeta(-3) = -\frac{1}{12} \cdot \frac{1}{10}$$

and this value actually has a physically meaningful interpretation in quantum physics that we will discuss in a later chapter.

The zeta function Euler used is well known in circles of people interested in prime numbers. Specifically, it is related to the distribution of prime numbers that is still a holy grail of mathematics and worthy of a brief side trip.

The German Riemann used the zeta function to make a hypothesis about the distribution of the prime numbers. He conjectured that the number of prime numbers less than a certain number n is related to the value of the zeta function at that number. But his conjecture still stands unproven.

So let us look at the relationship between prime numbers and twelve. We mentioned in the last chapter that twelve is certainly not a prime number, but it is related to a peculiar sequence of prime numbers. Intriguingly, twelve specific consecutive prime numbers

$11, 13, 17, 19, 23, 29, 31, 37, 41, 43, 47, 53$ sum up to the interesting number 364. This number is of course very close to the originally estimated number of days in a year as we have seen in the astronomical origins of our number and the number of accumulated gifts in the Christmas song. Strange coincidences, indeed.

Euler ingeniously used the zeta function, that resulted in the dumbfounding role of twelve in avoiding infinity, also to prove that there are infinitely many prime numbers. This statement on its face also may sound implausible. After all, if we have infinite number of integer numbers but only some of them are primes, how could there be infinitely many of them?

The simple answer is: because we can count them! We can simply map the prime numbers $2, 3, 5, 7, 11, \ldots$ to the sequence of the natural numbers $1, 2, 3, 4, 5, \ldots$ We can map the kth prime number to the kth natural number for any k value, no matter how large, all the way to infinity.

Well, this countable infinity (a concept invented by the German mathematician Cantor in the 19th century) may be a bit head-ache inducing and admittedly stretches the limits of our popular science level. Therefore, we will not ponder further implications of twelve's involvement with infinity.

10

Dozenal chemistry

It appears that nature is also a member of a dozenologists club of some sort, proven by the frequent occurrences of twelve in natural phenomena ranging from molecules to minerals.

There are natural chemical molecules exhibiting geometric manifestations of twelve. One of the most important molecules in human life and our world is carbon. Carbon's most stable configuration is carbon-12 containing 6 protons and 6 neutrons. It is prominently used in defining the dimensionless Avogadro number that specifies the number of atoms in a gram of any material. Avogadro's number is defined as the number of atoms in twelve grams of carbon-12 and it is $6.022 \cdot 10^{23}$, a fairly large number, but not in the league of the earlier very large universal numbers.

Carbon has other variations called isotopes; one of them is carbon-14 that is used in the archaeological technology of carbon dating and as such out of our interest here. However, there are other interesting carbon molecules in arrangements related to dodecahedron or icosahedron. They are somewhat new in chemistry having been discovered only in the second half of the 1980s. They do occur in nature and highly likely in outer space.

The simplest such carbon molecule is named C_{20} as it contains 20 carbon atoms. It is arranged in a do-decahedron, that as we have seen has 20 corners and 12 sides. The corners are the locations of the carbon atoms. A more common version is the C_{60} molecule, which is shaped as a spherical truncated icosahedron.

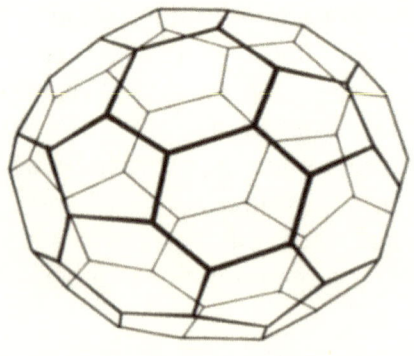

But, what is a truncated icosahedron? We can obtain one by "chopping off" the corners of a regular icosahedron that we are now intimately familiar with. The twelve truncated corners will be replaced by twelve pentagonal sides since there were 5 triangles connected to each corner. The twenty original triangular sides become hexagonal sides. The result of that blunt activity is bound to destroy its beauty, so we think. Well, apparently not as the image shows on the opposite page.

The truncated icosahedron is still very appealing with 90 edges, 60 corners and 32 sides. Noticeably the pentagonal sides do not touch each other directly, instead they are separated by the hexagonal sides. In case you wonder where you have seen that shape, it is the pattern of the modern soccer ball used around the world since 1970. It was first introduced in that year's World Cup in Mexico with alternating black and white pentagons and hexagons.

There are also further truncated icosahedron versions of the carbon molecule: C_{200} and even C_{540}. These are called icosahedral fullerenes. There are also some that encapsulate metals: they are referred to as metallo-fullerenes. Finally, the carbon-nanotubes are cylindrical fullerenes. There is also a class of fullerenes that have some of the carbon atoms replaced by nitrogen, called azafullerenes.

The carbon molecules are tiny things: the diameter of a C_{60} molecule is about 1.1 nanometer. The distance between the carbon atoms of the molecule is only 0.71 nanometer. This points to an interesting natural phenomenon involving twelve.

The preference of icosahedral shape in nanoscale is based on a balance between the surface forces of the individual atoms and the rest of the volume. Specifically, when the bonding force between atoms is strong, they tend to congregate with their twelve nearest neighbors. One of the most symmetric ways of doing this in three dimensional space is by the so-called icosahedral clustering. Small clusters of noble gases and some metal atoms also behave this way.

The limit of such clustering is due to the fact that the three dimensional space cannot be filled voidlessly by icosahedra. Hence when the gathering gets too large, the volume forces override the inter-atom forces. The result is that the atoms on the exterior of the cluster rearrange themselves into a cubic shape.

A very important molecule, related to carbon and exhibiting a numerical preference for twelve is the benzene. Its chemical composition is C_6H_6 and is organized as an inner ring of six carbon atoms and a single hydrogen atom connected to each of them. It is a highly unsaturated molecule having only a single hydrogen for each carbon atom, since carbon could tie several more. This, however, enables the high level of cyclic connectivity of benzene molecules.

Benzene is derived from petroleum and unhealthy for humans in any form. It is the fundamental component of many practical materials of our everyday lives, ranging from industrial plastics to nylon textiles.

There are other carbon related molecules that also prefer twelve atoms, such as $C_{10}N_2$. It is a member of the family of carbon nitrides, specifically the group of dicyanopolyynes that are composed of a chain of carbon atoms ending in nitrogen atoms. The chain itself is connected via alternating single and triple bonds. The dicyano is the reference to the two nitrogen atoms.

A variation of this is the compound $HC_{10}N$ where one of the nitrogen atoms at the end is replaced by a hydrogen atom. Another combination is the CH_3C_7N molecule where one end is completed with three hydrogen atoms connected to the last carbon atom. Note, that there are still twelve atoms. The description of the properties of these compounds is beyond our scope, let them just be testimonials to nature's relentless use of the number twelve.

Other, not carbon related chemicals favoring twelve atoms in their molecules also exist. A variation of boron, called amorphous boron, B_{12}, is also in the shape of an icosahedron. These molecules are regionally bonding depending on their neighborhood and usually in random fashion. It is a naturally occurring component of meteorites. Apparently extraterrestrial nature also likes twelve.

The chemical composition of the pyrite is iron-sulfide with the formula FeS_2. As a bi-product of iron mining, pyrite has been known by humans for millennia. Its practical applications range from being the ignition component of old firearms to use in modern batteries. As an igniter of firearms it was struck by the cock to create a spark and in modern use it is the cathode in lithium batteries.

Most naturally occurring pyrites are of the shape of a cube, but there is a dodecahedron shaped exception by nature. These are not regular dodecahedra because the twelve sides are not regular pentagons albeit still pentagonal in shape. As a consequence, the edges are not the same length either, 24 of them are the same and another 6 are of different length. This modification results in losing some of the many symmetries of the dodecahedron. Due to its shiny brass coloring, miners often mistook the dodecahedral pyrites for gold resulting in the name "fool's gold".

There are also more complex minerals, called quasicrystals. Quasicrystals are solid materials that use a quite unique organizing principle for arranging their atoms. They usually, but not necessarily contain a central atom that is surrounded by an icosahedral shell of 12 atoms that are located in the corners. They may also contain a second shell of 32 atoms where in addition to the corners, atoms are also located in the centers of the 20 sides. Finally a third shell may also exist of 62 atoms where the additional atoms are located in the middle of the 30 edges.

Originally quasicrystals were thought to be physically impossible, then some were produced under delicate laboratory environments. The simplest of these produced were the icosahedral quasicrystals Al_6Mn and Al_6CuLi_3, both with aluminum, and one with manganese while the other with copper and lithium.

Their atomic structure does not have translational symmetry, so they are not repetitive in a certain coordinate direction as regular crystals are. Quasicrystals, however, have rotational symmetry along certain axes with 2, 3, or even 5-fold symmetries, the latter type never present in regular crystals.

Finally, a quasicrystal was also discovered in nature and called the icosahedrite. It is a metallic grey, somewhat brittle material with chemical composition $Al_{63}Cu_{24}Fe_{13}$, in other words composed of aluminum, copper and iron isotopes. It was found in a rock sample gathered in the Kamchatka territory of Russia, near the Khatyrka river. Since it was found in a remote and isolated region, it was not considered to be man made and was approved by the Commission of New Minerals and Mineral Classification as a natural mineral in 2010.

The sample actually demonstrated a combination of six intersecting 5-fold axes that is characteristic of the icosahedron, hence its name. One cannot but wonder about the connection to the 5-transitive M_{12} symmetry group. It also exhibited three-fold and two-fold symmetries found by X-ray diffraction techniques that produced symmetric diffraction patterns representative of such symmetries.

The uniqueness and rareness of the icosahedrite instigated the hypothesis that it is extraterrestrial in origin and arrived to Earth by way of an asteroid. The icosahedral shell structure of quasicrystals resembles the generation of the dodecahedral numbers presented in a prior chapter, and since several quasicrystals were now created in laboratories, it appears that we are simply following nature's footsteps.

11

Nature's twelve

You may ask whether nature's use of twelve with its related geometric symmetries is confined to the non-living chemical world, or also appears in living organisms. Well, the answer is yes, it does. Certain living organisms also like twelve. For example, we humans have twelve cranial nerves and have a twelve inch long part of the small intestine, called duodenum, although this might be coincidental since the duodenum measured in centimeters is not twelve.

Some microorganisms use a regular polyhedral arrangement of repeating proteins because it is spatially efficient and also minimizes the genetic information necessary to recreate the organism. It is also hypothesized that the icosahedral arrangement minimizes energy when interacting particles are located on the surface of a sphere.

Nature's recognition of that fact is apparently original, evidenced by the structure of one of the very first eukaryote life forms, the radiolaria, evolved about 550 million years ago. They were floating in the ocean currents, fed themselves by filtering water and capturing small organisms. The radiolaria had silica shells of intricate Platonic solid shapes and since their silica shell is not dissolved by sea water, many fossils have been found.

This richness of fossil records and forms drew the attention of the turn of the 20th century German biologist, Häckel. He published a book titled "Artforms of nature" containing a large collection of radiolarian images. He was a talented illustrator and some of his drawings in reprints are now cherished by collectors. Among them are the Circugania icosahedra, shown below, with a distinctly icosahedral shape.

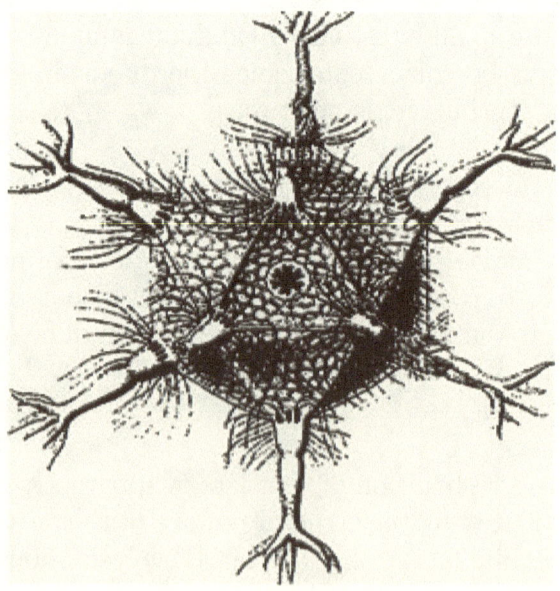

Since then other living creatures have been recognized to have icosahedral structures, such as the chicken-

pox, hepatitis and the infamous, but now defeated, po-
lio virus. But the most widespread icosahedral virus
family, the adenovirus, still affects our everyday lives
unfortunately. The name originates in the adenoid
gland where they were first identified. There are 57
different variations of them found in humans.

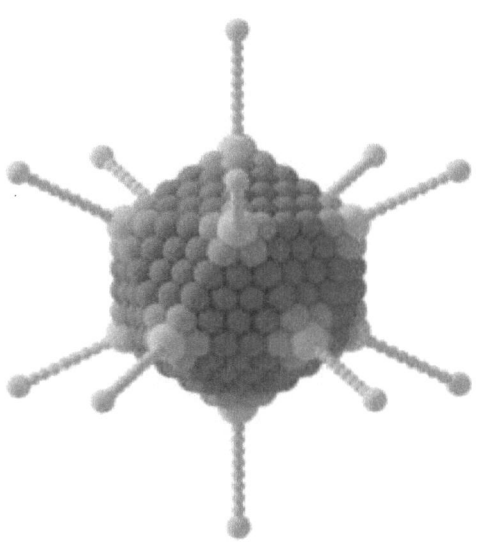

The adenovirus, shown above, has a protein enve-
lope that is icosahedron shaped with a diameter of
about 100 nanometer or 0.1 micrometer, rather large
as far as the viruses are concerned. It is comprised of
spherical component nodes that are identical in con-

tent but different in role depending on their position in the structure.

For example, each corner node of the icosahedral protein structure has a spike. These external spikes are used by the virus to attach itself to the cell it invades. The virus itself contains about a million amino acids, but still considered to be a rather simple virus, again by virus standards.

There are 252 little spherical nodes on the surface of the virus, in a way resembling the icosahedral shells of the quasicrystals with a larger number of side and edge nodes. There are 21 nodes on each side. From those 6 are inside the sides, 12 are on the edges and 3 are on the corners. Since there are 20 sides, 20 times 6 gives 120 nodes of them. Similarly, the 20 times the 12 edge nodes would give 240, however, they are shared between two sides, hence there are only 120 distinct edge nodes. Finally, there are simply 12 distinct corner nodes corresponding to the corners of the icosahedron. Hence the total number of surface nodes is $252 = 120+120+12$.

The way the virus works inside us is far beyond our mathematical focus, suffice to say that it can cause nasty infections of the upper respiratory system. Tonsillitis, ear infection and even pneumonia could be the result of adenovirus. We might question whether the efficiency of the icosahedron hiding in its beauty is worthy of our admiration at all. But we cannot deny nature's tenacity regarding the icosahedron, its use 550 million years ago and its use today.

We now abandon the geometrical manifestations of twelve and look at its algebraic presence that enables us to go even smaller in scale. What could be smaller than viruses? Well, elementary particles of matter, of course. And these are truly nature's twelve, since there are twelve of them.

This was not always apparent, it took us humans millennia to find them all. First, in ancient times we thought that there was just a single particle, the atom. Then about a hundred years ago the atom was broken, first into two particles, the electron and the proton. Then came the neutron that was soon also broken and we found the quarks with their up and down versions.

This up and down pair hints at some very specific symmetry considerations that are abundant in nature, specifically mirror symmetry. We discussed rotational symmetry at length in earlier chapters, but mirror symmetry has its own strength. Any particle that is mirror symmetric cannot be turned into its pair by rotations. The example often quoted is a pair of left and right shoes which perfectly manifest this scenario. They look identical in the mirror but can never be made the same.

As we know now, there are three pairs of quarks: up and down, top and bottom, and the strangely named pair of charm and strange. The group of leptons also contains three pairs or particles: the electron and electron neutrino, the muon and its neutrino, the tau and its neutrino. These six pairs then comprise twelve particles of matter in general.

There is another hidden layer of the use of number twelve inside these twelve particles. First, the formula developed by physicists for the mean lifetime of the muon, while it is too complex to show here, contains a multiplier of 12 in it. Secondly, the formula for the mean lifetime of the neutron has an exponent of 12 in it. Thirdly, the formula for the mean lifetime of the proton has both a multiplier and an exponent of 12 in it. Rather conspicuous occurrences of the number twelve.

Then the formula for the particle masses contains an exponent of 12 in two distinct positions and the formula computing the electron spin factor also has an exponent of 12. Finally, the value of the so-called Casimir force originally proposed by the Dutch physicist in the middle of the 20th century is computed by a formula around a $\zeta(-3)$ term we discussed earlier and whose value also contained 12. The force results from the atomic level interaction between two plates in vacuum, hence it appears that the numerical twelve has a deeper presence in particle land than just being the number of the particles.

Since we are talking about a force we should note that there are the four types of particles carrying forces: photons (carrier of the electromagnetic force), gluons (producing the strong nuclear forces), bosons (sources of the weak electromagnetic forces) and the hypothetical gravitons (the medium of the gravitational force). Since these forces in our three dimensional physical space all have three components along the spatial directions, we also have twelve fundamental force components of nature.

The boson class became complete with the recent finding of the Higgs boson besides the Z and W bosons. It appears, and scientists concur, that we have now discovered them all. Higgs boson accounts for the intrinsic mass of the particles. At its discovery it was dubbed by the name of the God particle. Well, we are still searching for the graviton particle, whose effects are undeniable but form is illusive.

Maxwell unified the electric and magnetic forces a century ago and physicists have been able to unify the weak and strong nuclear forces. They are now searching for a Grand Unified Theory of particle physics that would combine them in one single force. One of the current proposals is the F-theory which is a 12-dimensional string theory with ten spatial and two temporal dimensions. This is as hard to visualize as controversial, so we will just view in wonderment of twelve striking again.

Finally, we are also still searching for the unification theory of all natural forces: the theory of everything. However, after a lifetime of research even Einstein was unsuccessful in unifying the nuclear and electromagnetic forces with gravity. Despite ongoing efforts by many scientists we are still unable to do so. This might be accomplished in the future once we understood the hypothesized dark matter about which we are very much in the dark except suspecting its presence.

12

Corners of the Earth

We are now ready to find the corners of the Earth indicated in the subtitle of this book and depicted on the cover picture. Projecting the image of the Earth onto a sheet of paper (or papyrus or fabric in ancient times, as in the case of Eratosthenes' map) had been attempted in many forms and most were fraught with distortions of the known geographic entities, continents, islands, lakes and oceans, especially the latter due to their vast sizes.

The most widely known are the familiar maps with the Mercator projection that relies on projecting the great circles of the spherical globe, longitudes and latitudes, onto a plane. In this projection the circles of longitude and latitude become straight parallel lines. While this projection fits our mental vision of Earth, certain areas are extremely distorted at higher latitudes producing unrealistic images of the geography in the polar regions.

One cannot but recognize the icosahedron shape's close proximity to a sphere defined by the corners of it. Such recognition led R. Buckminster Fuller in the early 1940's to create a map of our planet based on an icosahedron inscribed inside the spherical Earth. Fuller assigned the poles inside a triangular side of the icosahedron but not at a corner of it. The corners were at special latitudes and longitudes. He projected each triangle edge separately to a partial great circle, as shown by the red yarn on the globe of the cover image.

We have seen in an earlier chapter that the edge of the regular icosahedron inscribed into a sphere is approximately 1.05 times the radius of the sphere. Hence, the twenty edges of the icosahedron form equilateral triangles with sides approximately the radius of the Earth and the appearance of the icosahedron shaped Earth was very globe-like.

To further minimize the distortion of the projection, Fuller placed the corners of these triangles not exactly on the surface of the Earth, but somewhat below the surface. This selection significantly enhanced the graphical fidelity of the geographic details of the Earth. The relative sizes of measured earthly entities were largely retained in Fuller's projection.

Fuller then decided to present this three dimensional high fidelity map in 2 dimensions by cutting up this icosahedron shaped Earth along certain edges. In this arrangement the continents appear to be an almost contiguous land mass, sometimes called the Earth island. He called his map the Airocean World map or the Dymaxion map.

The word Dymaxion, coined by Fuller himself and by now present in encyclopedia, means the concept of maximizing the results of some technological solution while minimizing the material and energy needed to accomplish the goal.

Fuller's map was first published in the early 1940's and reproduced in many publications since then. The image below was created by Eric Gaba, Wikimedia Commons user by the name of Sting.

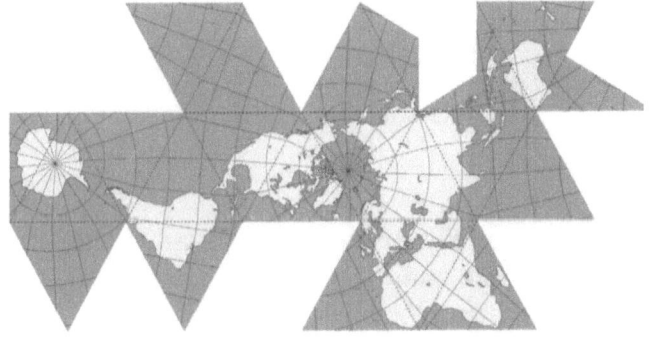

The planar layout, while largely rather faithful to geographic relations, cut through two geographical entities that should have been kept intact. Fuller corrected this by deviating from the layout of the original 20 triangles in the plane. He strategically cut two triangles and placed them in a different place.

The leftmost triangle in the middle row was cut in half and moved to the right hand side of the top row to avoid the southern tip of Australia being cut off from the rest. Another triangle, the second from the right on the top row, was cut at an angle and rotated counterclockwise to the left to reconnect the lower tip of the Japanese island with its mainland. Also peculiar is the equator on Fuller's map in two segments in contrast to our familiar single horizontal line.

On the other hand, the map's higher fidelity in the polar regions compared to the Mercator map is notable. The middle row leftmost triangles render a very accurate contiguous image of Antarctica and the same could be said about the fidelity of the northern polar region. Nevertheless, the map never caught on in the industry and is now largely forgotten by the everyday population.

The map, when creasing and folding at the triangular edges then gluing the exterior edges together, could be used to recreate the 3 dimensional icosahedron shaped Earth. As such it became known as Fuller's Dymaxian Globe^{TM}, a popular toy for children, and was used even in a board game named World Game.

The process of projecting the earthly features to the triangular sides of the icosahedron may be generalized. Let us assume that the edges of the planar triangles of an icosahedron inscribed into a sphere are subdivided into 2, 3 or even 4 segments. Then each triangle will be covered by 4, 9 or 16 smaller triangles that are still equilateral. Projecting the corners of these triangles outward onto the surface of the sphere, we get a so-called tessellation of the sphere. The subdivision by 2 segments will result in a tessellated sphere with 80 facets, the next one with 180 and the final one with 320 facets. The three versions of the upper hemisphere of the tessellated sphere are shown in the figure below.

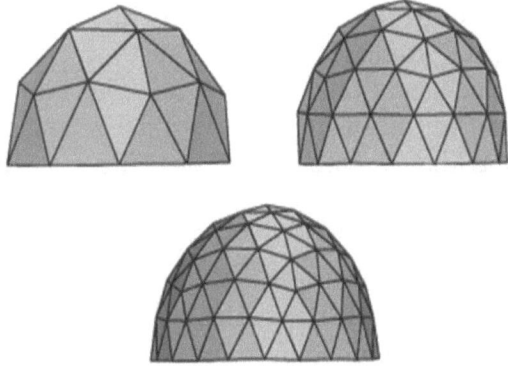

Each of the edges of the triangles is visible in the same fixed angle when viewed from the center. Using the aforementioned ratio of the edge of the icosahedron to the radius, we find the angle to be 63 degrees, 26 minutes and about 5 seconds for the regular icosahedron. This is very close to the 60 degree angles of the triangles themselves, hence these volume sectors of the icosahedron are very proportional.

This angle is simply divided by 2, 3, or 4 as the edges are subdivided. Using the above formula, we can then obtain the edges of the twice, thrice and quadruple tessellated sphere as 0.5464, 0.3670, and 0.2759. The edges and the angles become smaller and smaller as the sides are subdivided. After a certain number of subdivisions, the tessellated sphere can barely be distinguished from the regular sphere.

Fuller recognized the practical importance of this geometrical phenomenon and invented the geodesic dome. This is in essence the continuation of the above process to a higher level. He called his method triangular geodesic transformational projection. The important aspect of the invention is that the final triangles are still planar triangles and, as such, enable easy industrial production.

The engineering robustness of the geodesic dome construction can be proven theoretically and validated by the numerous stadium roofs built with this concept. Even the famous New York Ball on Time Square is a geodesic dome, an almost complete sphere.

At one point geodesic domes were considered to be a major solution to affordable housing development using prefabricated steel bar triangle components. While such structures still retain their practical importance in the form of easily assembled temporary housing tents for people displaced by some natural disaster, the original enthusiasm about them has now fizzled.

Geodesic structures still appear in agricultural green houses constructed from aluminum frames of geodesic arrangement and simply covered with plastic sheets. They also still occur in constructions where large sized building materials are difficult to deliver, for example in mountainous regions.

Fuller did not stop at stationary objects. He also applied his invention to building cars. His prototype car, the Dymaxion car, was an asymmetrical geodesic dome to lessen the aerodynamic drag. It was expected to reach 30 miles per gallon at the time when most cars achieved at most 10. It was to reach a top speed of 90 miles per hour, another respectable number at the time.

It was also designed to be easily assembled from the geodesic dome components and, even more, could be disassembled, an idea unheard of until then. Due to the narrower rear to create an aerodynamic profile, the rear axle contained only one wheel. This made the car difficult to steer, especially corner with, to the extent that it resulted in serious accidents. One of the original three prototypes was demonstrated at the 1933 World Fair with a dubious result when it overturned and fatally injured its driver.

Despite the lesser industrial and practical success of his invention Fuller did not lose his belief in the topic. He had some, then rather grandiose, now one can say futuristic, ideas about covering whole cities with geodesic roofs thereby shielding humans from nature's elements. He actually considered this more like redirecting as opposed to shielding, because he proposed to gather the rain fallen on the dome and to be used for irrigation.

He was always thinking about the possibilities of
"doing more with less". He even extended his geometric thinking into the philosophical realm and wrote
about "geodesic life". The way he expressed this was
that "he wanted to reorganize the environment of man
in a way by which greater numbers of them can pros-
per". This is an avenue we are not going to follow
here, but only mention with respect.

Fuller's contribution to society was ultimately ac-
knowledged and his name immortalized by the fullerene
molecules described in a prior chapter. All those were
named after Buckminster Fuller, because of their re-
semblance to the geodesic dome and they are sometimes even called Bucky-balls.

Using a highly tessellated sphere to create a more detailed Fuller projection map would provide us with a
geodesic planet with many corners of our Earth. But
the projection does not have to end at the Earth's
surface. Imagine extending the radii pointing to the
corners of the tessellation triangles toward the skies.
These will point to a certain sector of the celestial
sphere at certain astronomical distance.

Of course, in space we cannot confine ourselves to
a spherical surface only. In order to still utilize our
icosahedron based tessellated view of the universe we
can consider the volumes defined by a single facet of
the tessellation from our Earth-centric vantage point.
These volumes are tetrahedra with an external side on
the tessellated surface and three internal sides adjacent to the edges of the external triangle.

These tetrahedra fill in the volume of the sphere (or the universe) to a certain extent depending on the granularity of tessellation. We saw earlier that the original regular icosahedron's 20 tetrahedra occupy about 60.5 % of the volume. Increasing the tessellation by 2, 3, or 4 as before, the tetrahedra will occupy 87.3, 94.1, and 96.6 % of the tessellated universe.

Further levels of tessellations could enable us to see more details of our Milky Way than our ancestors saw looking into the sky millennia ago. We could see the fascinating spiral arms that are approximately following a mathematical curve called logarithmic spiral. The curve was first described by the French Descartes, but the Swiss Jacob Bernoulli made it very famous by calling it the "wonderful spiral". The main reason for his enthusiasm was the fact that the shape of the spiral is self-similar, any portion of it looks like any other portion apart from being rotated.

The curve occurs frequently in nature ranging from the interior of nautilus shells through patterns of cyclones to the shape of spiral galaxies. The amazing fact is that the spiral arms of our Milky Way galaxy are pitched (the angle between the tangent of the curve and the tangent of the circle going though the same point) by a constant twelve degrees, give or take a half a degree. This angle is unique among the many spiral galaxies found in the estimated $2.7 \cdot 10^{23}$ mile diameter universe, a thought worthy of closing a book about the number twelve.

Epilogue

The reasons for the human penchant for the number twelve appears to be understood although we are still trying to gather our minds around the mind boggling appearance of the number 12 in the sum of an infinite sequence. Similarly, the quest is ongoing to fully understand nature's bias toward twelve components, symmetries of twelve and dodecahedral, icosahedral shapes.

We can ask: is our geometry describing nature, or is nature mimicking our geometry? We can ponder: is mathematics invented by us humans or do we just discover pieces of it as we go along? If it is the latter case; who embedded the mathematical phenomena originally into nature?

Without lapsing into further speculations, it is left to the reader to attempt to find answers that are personally satisfying. But always remember: unknown is not necessarily unlikely.

References

Barrow, John D.: The constants of nature, Random House, 2003

Conway, John and Sloane, Neil: Sphere packings, Lattices and Groups, 3rd edition, Springer, 1998

Darling, David: The universal book of mathematics, From abracadabra to Zeno's paradoxes, Wiley, 2004

Derbyshire, John.: Prime obsession: Bernhard Riemann and the greatest unsolved problem in mathematics, J. Henry Press, 2003

Fuller, R. Buckminster and Applewhite, E. J.: Synergetics, explorations in the geometry of thinking, Mac millan Publishing, 1975

Gray, Robert W.: Fuller's Dymaxion^{TM} map, Cartography and Geographic Information Systems, 21(4): 243-246, 1994

Keyes, Gene: Evolution of the Dymaxion map; An illustrated tour and critique. www.genekeyes.com

Nystrom, J. W.: Duodenal system of arithmetic,

measures, weights and coins, Porter & Coates, 1875. Reprinted by the Dozenal Society of America

Rees, Martin: Just six numbers; The deep forces that shape the universe, Basic Books, 2000

Ronan, Mark: Symmetry and the Monster, One of the greatest quests in mathematics, Oxford University Press, 2006

Stewart, Ian: Nature's numbers; The unreal reality of mathematics, Basic Books, 1995

Wells, H. G.: The outline of history, 1920, Reprinted in 1956, Garden City Books